Bilingual edition

精密計測学

高　　偉
清水裕樹
水谷康弘
道畑正岐
河野大輔
吉田一朗
伊東　聡
清水浩貴 [著]

Precision Metrology

Wei Gao
Yuki Shimizu
Yasuhiro Mizutani
Masaki Michihata
Daisuke Kono
Ichiro Yoshida
So Ito
Hiroki Shimizu

朝倉書店

Bilingual edition

Seimitsu Keisokugaku
Precision Metrology

Wei Gao, Yuki Shimizu, Yasuhiro Mizutani, Masaki Michihata,

Daisuke Kono, Ichiro Yoshida, So Ito, Hiroki Shimizu

ISBN 978-4-254-20178-9

まえがき

　精密計測学は物体の形と動きの計測を中心に扱う学問であり，科学技術全般および製造現場を支える重要な基盤学術分野である一方，最新の科学技術を取り入れながら進化し，常にその時代の科学技術の最先端を走り続けている．精密計測にとって，精度向上が永遠のテーマであり，また製造現場においては，精密計測の高効率化とロバスト性も強く求められる．一方，関連技術の急速な発展に伴い，各種精密測定機の性能が格段に向上してきている．また，自律校正法に代表されるような知的精密計測の手法も種々開発され，さらに人工知能（AI）と機械学習を取り入れることで，精密計測の分野で新たな可能性を見出している．

　本書では最先端精密計測技術を取り上げながら，アッベの原理など，その基本となる精密計測学の原理原則に重点を置き，焦点を絞りながら丁寧に記述している．第1章では精密計測の基準となる国際単位系（SI）と計測標準について，第2〜4章では，精密計測の基本技術としての長さスケール，角度スケール，時間スケールについて，第5〜10章では，精密計測の応用技術としての幾何形状と表面性状の計測，光干渉計，マシンビジョン，空間位置計測，光学顕微鏡，走査プローブ顕微鏡について，第11〜14章ではデータ処理に着目して，誤差要因と不確かさ，自律校正法，機械学習と精密計測，超短パルスレーザと光周波数コムについてそれぞれ記している．

　本書は姉妹編の『Bilingual edition 計測工学 Measurement and Instrumentation』（朝倉書店）と同様に，図，表および式がページの中央に，上下にはそれぞれ和文と英文の記述が対応するように配置し，効率的かつ明快なページレイアウトにしている．日本語と英語の両方で精密計測学が学べる教科書と参考書として，学部および大学院学生と技術者の一助となれば幸いである．

　本書の出版にあたり，産業技術総合研究所 高辻利之先生，稲場肇先生，尾藤洋一先生，東北大学 山口健先生，鈴木研先生から貴重な修正意見を頂いたこと，大阪大学 高谷裕浩先生，東京大学 高橋哲先生より多大なご支援を頂いたこと，東北大学 松隈啓先生，佐藤遼先生および学生諸君から原稿の校正に御協力を頂いたこと，また朝倉書店編集部に御尽力を賜ったことを記し，心から感謝申し上げる．

2024 年 1 月

著者代表　高　偉

Preface

Precision metrology is an academic field that deals with the measurement of dimensions and motions of an object. It is one of the corner stones to support the innovations in science and technology as well as the activities in manufacturing industries. Significant progress has been achieved in precision metrology with the advancement in related technological areas. Meanwhile, innovations have been made in intelligent precision metrology represented by self-calibration. New opportunities are also being created in precision metrology by implementing artificial intelligence (AI) and machine learning technologies.

In this book, the mathematical and physical principles behind the newest technologies of precision metrology, such as the Abbe principle, are focused and clearly explained. Chapter 1 presents the standards of precision metrology including the new SI. As basic technologies, length scale, angle scale and time scale are descried in Chapters 2 to 4, respectively. As integrated technologies, measurement of geometrical form and surface texture, interferometry, machine vision, measurement of volumetric position, optical microscopy, scanning probe microscopy are addressed in Chapters 5 to 10, respectively. Chapters 11 to 14 focus on data processing, with the topics of error sources and measurement uncertainty, self-calibration methods, machine learning and precision metrology, ultrashort pulse laser and optical frequency comb.

This book adopts an efficient and easy-to-read page layout by locating the figures, tables and equations in the middle of the page sandwiched by Japanese and English texts, which is the same as its sister book "Measurement and Instrumentation" (Bilingual edition, Asakura Publishing). It can be used as a textbook or a reference book by either Japanese or English speaking students in both undergraduate and graduate courses and engineers for learning precision metrology in the two languages.

We would like to thank Dr. Toshiyuki Takatsuji, Dr. Hajime Inaba and Dr. Youichi Bitou of AIST, Prof. Takeshi Yamaguchi and Prof. Ken Suzuki of Tohoku University for providing invaluable suggestions to some of the chapters, Prof. Yasuhiro Takaya of Osaka University and Prof. Satoru Takahashi of The University of Tokyo for their kind support to this book, Prof. Hiraku Matsukuma, Prof. Ryo Sato and the students of Tohoku University for proof-reading the manuscript. The dedicated efforts from Asakura Publishing are also appreciated.

January, 2024

Wei Gao, on behalf of the authors

目　次
Table of Contents

執 筆 者
Authors

高　偉　Wei Gao　　　　　　　　　　　　　(Chapters 1, 11, 12, 14)
東北大学大学院工学研究科教授
Professor in Graduate School of Engineering,
Tohoku University
Fellow, The Engineering Academy of Japan

清水裕樹　Yuki Shimizu　　　　　　　　　　(Chapters 2, 3)
北海道大学大学院工学研究院教授
Professor in Graduate School of Engineering,
Hokkaido University

水谷康弘　Yasuhiro Mizutani　　　　　　　　(Chapters 4, 13)
大阪大学大学院工学研究科准教授
Associate Professor in Graduate School of Engineering,
Osaka University

道畑正岐　Masaki Michihata　　　　　　　　(Chapters 6, 9)
東京大学大学院工学系研究科准教授
Associate Professor in Graduate School of Engineering,
The University of Tokyo

河野大輔　Daisuke Kono　　　　　　　　　　(Chapters 7, 8)
京都大学大学院工学研究科准教授
Associate Professor in Graduate School of Engineering,
Kyoto University

吉田一朗　Ichiro Yoshida　　　　　　　　　　(Chapter 5)
法政大学大学院理工学研究科教授
Professor in Graduate School of Science and Engineering,
Hosei University

伊東　聡　So Ito　　　　　　　　　　　　　(Chapter 10)
富山県立大学工学部准教授
Associate Professor in Faculty of Engineering,
Toyama Prefectural University

清水浩貴　Hiroki Shimizu　　　　　　　　　　(Chapter 11)
九州工業大学大学院工学研究院准教授
Associate Professor in Graduate School of Engineering,
Kyushu Institute of Technology

第 1 章　精密計測の基準

　精密工学の基盤分野の 1 つとして，精密計測は長さを基本とする物体の形と動きの計測を中心に扱う学問である．測定対象となる量は多様な材料と形状を持つ加工物や機械の寸法，表面形状，表面粗さ，位置，速度および加速度など多岐にわたるため，その対象に適した測定機器と測定法を選んで測定を適切に実施する必要がある一方，測定結果の信頼性を示す精度あるいは不確かさは精密計測において最も重要な位置づけであり，基準となる測定標準による切れ目のない校正の連鎖によって，すべての測定機器と測定結果を国際単位系（SI）および国家標準にトレーサブルとする必要がある．本章では，SI 単位系とトレーサビリティも含めて，精密計測の基準について記述する．

Chapter 1　Standards of Precision Metrology

　Precision metrology, as one of the disciplines within precision engineering, is an academic field that focuses on the dimensions and motions of objects. This field encompasses a wide range of length-based target quantities, known as measurands. These measurands include, for example, dimensions, surface forms, surface finishes, positions, velocities, and accelerations of workpieces and machines made of different materials and in different shapes. Achieving accurate measurements in precision metrology necessitates careful selection of appropriate measuring instruments and measurement methods. In addition, a crucial issue within precision metrology is determining the accuracy or uncertainty that demonstrates the reliability of the measurement results. This reliability is ensured through traceability to the International System of Units (SI) and corresponding national standards, accomplished via a continuous chain of calibration using measurement standards. In this chapter, the measurement standards, SI, and traceability for precision metrology are presented.

● 1.1　国際単位系（SI）

SI 単位は 7 つの基本量/基本単位，基本量/基本単位を組み合わせた組立量/組立単位，および 24 の接頭語で表現される．7 つの基本量/基本単位は，時間/秒，長さ/メートル，質量/キログラム，電流/アンペア，熱力学温度/ケルビン，物質量/モル，光度/カンデラとなっている．SI 基本単位の相互関係を Fig. 1-1 に示す．各 SI 基本単位の定義に用いられる基礎物理定数も図に記載されている．これらの基礎物理定数の値は不確かさが 0 となっている．このように，すべての SI 基本単位は普遍的な基礎物理定数を用いて自然界の法則に従って定義されており，特定の物質に依存せず，高安定で高精度な特徴がある．一方，図に示すとおり，秒とモル以外の基本単位は単独では決

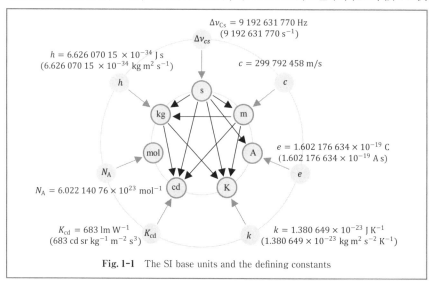

Fig. 1-1　The SI base units and the defining constants

● 1.1　The International System of Units（SI）

The International System of Units (SI) comprises seven base quantities/units, derived quantities/units from the base quantities/units, and twenty prefixes. These base quantities/units are time/second, length/metre, mass/kilogram, electric current/ampere, thermodynamic temperature/kelvin, amount of substance/mole, and luminous intensity/candela. Fig. 1-1 shows the relationship between the seven SI base units. This figure also illustrates the fundamental physical constants used to define the SI base units. The numerical values of the seven defining constants have zero uncertainty. As the SI base units derive their definitions from the laws of nature and are grounded in physical constants rather than human artifacts, they are stable and accurate. Meanwhile, the definitions of the SI base units, except for the

定されず，他の基本単位にも依存する．例えば，メートルとアンペアは秒に依存し，キログラムは秒，メートルに依存している．また，モル以外の基本単位は，すべて秒に依存する．精密計測に密接に関連する SI 基本単位の秒，メートル，ケルビンはそれぞれ以下のように定義される．

■秒（記号は s）は，時間の SI 単位であり，セシウム周波数 $\Delta\nu_{Cs}$，すなわち，セシウム 133 原子の摂動を受けない基底状態の超微細構造遷移周波数を単位 Hz（s^{-1} に等しい）で表したときに，その数値を 9 192 631 770 と定めることによって定義される．

■メートル（記号は m）は長さの SI 単位であり，真空中の光の速さ c を単位 $m\,s^{-1}$ で表したときに，その数値を 299 792 458 と定めることによって定義される．ここで，秒はセシウム周波数 $\Delta\nu_{Cs}$ によって定義される．

■ケルビン（記号は K）は，熱力学温度の SI 単位であり，ボルツマン定数 k を単位 $J\,K^{-1}$（$kg\,m^2\,s^{-2}\,K^{-1}$ に等しい）で表したときに，その数値を $1.380\,649\times10^{-23}$ と定めることによって定義される．ここで，キログラム，メートルおよび秒は h, c および $\Delta\nu_{Cs}$ に関連して定義される．

second and mole, rely on that of others. For instance, the definitions of the metre and ampere rely on that of the second, while the definition of the kilogram relies on that of the second and metre. Furthermore, all the SI base units, except the mole, rely on the second. The SI base units (second, metre and kelvin) with widespread use in precision metrology are defined as follows:

■The second, symbol s, is the SI unit of time. It is defined by taking the fixed numerical value of the cesium frequency, $\Delta\nu_{Cs}$, the unperturbed ground-state hyperfine transition frequency of the cesium 133 atom, to be 9 192 631 770 when expressed in the unit Hz, which is equal to s^{-1}.

■The metre, symbol m, is the SI unit of length. It is defined by taking the fixed numerical value of the speed of light in vacuum, c, to be 299 792 458 when expressed in the unit $m\,s^{-1}$, where the second is defined in terms of the caesium frequency $\Delta\nu_{Cs}$.

■The kelvin, symbol K, is the SI unit of thermodynamic temperature. It is defined by taking the fixed numerical value of the Boltzmann constant, k, to be $1.380\,649\times10^{-23}$ when expressed in the unit $J\,K^{-1}$, which is equal to $kg\,m^2\,s^{-2}\,K^{-1}$, where the kilogram, metre and second are defined in terms of h, c and $\Delta\nu_{Cs}$.

● 1.2　校正とトレーサビリティ

測定機器を使って正確に計測を行うためには，その測定機器はより正確な（不確か
さがより小さい）標準器によって校正されることが必要条件である．この標準器もよ
り正確な標準器によって校正される，というようにより正確な標準器を求めていくと，
SI単位が実現される国家標準に辿り着く．測定機器が校正の連鎖によって国家標準に
辿り着けることが確かめられている場合，この測定機器により得られた結果は国家標
準にトレーサブルであるという．トレーサビリティとは，「不確かさがすべて表記され
た切れ目のない比較の連鎖によって，決められた基準に結び付けられうる測定結果ま
たは標準の値の性質」である．ここで，基準は通常SI単位，またはSI単位が実現さ
れる国家計量標準である．常用参照標準，実用標準などの測定標準も基準となりうる．

日本では計量法に基づき，経済産業大臣が特定標準器あるいは特定標準物質を国家
計量標準（特定標準器）として定め，校正とトレーサビリティの源となるものにして
いる．Fig. 1-2に国家標準の供給制度と校正事業者の登録制度として導入されている
計量法トレーサビリティ制度（JCSS）の概要を示す．特定標準器は産業技術総合研究

● 1.2　Calibration and traceability

Calibrating a measuring instrument using a measurement standard of high accuracy is essential for obtaining reliable measurements using the instrument. For re-
validation, calibrating the measurement standard using a different standard with
superior accuracy is necessary. This procedure is repeated until calibration is performed according to a national standard, establishing that the measurement result
obtained from such an instrument can be confidently traced back to the national
standard. Metrological traceability is defined as the property of a measurement result whereby the result can be related to a reference through a documented unbro-
ken chain of calibrations, each contributing to the measurement uncertainty. The
reference can be considered to be the definition of a unit of the SI, through its practical realization. The reference could also be any measurement standard for a quantity of the same kind.

In Japan, the Ministry of Economy, Trade and Industry (METI) designates the
national primary standards or reference materials as the sources of calibration and
traceability. These standards are recognized under the Measurement Act as national measurement standards. Fig. 1-2 shows a schematic of the Japan Calibration Service System (JCSS), which contains the National Standards Provision System and
the Calibration Laboratory Accreditation System. The national primary standards

所, 日本電気計器検定所または経済産業大臣が指定した指定校正機関が運用し, 特定二次標準器の校正に用いられる. そして, 事業者登録制度にて登録された登録事業者は特定二次標準器を運用し, 常用参照標準の校正を行う. さらに, 次の階層にある登録事業者はその常用参照標準を保有し, 一般ユーザが精密計測などで用いられる実用標準や一般測定機器を校正する. それによって, 精密計測などの測定結果が国家標準および国際単位系へ関連付けられるようになる. なお JCSS に基づいて行われた校正に対して, JCSS 認定シンボルの入った校正証明書が発行され, 日本の国家計量標準へのトレーサビリティが示される.

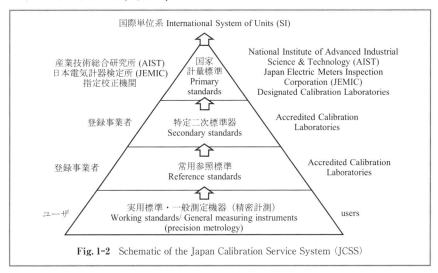

Fig. 1-2 Schematic of the Japan Calibration Service System (JCSS)

are established and operated in the National Metrology Institute of Japan/National Institute of Advanced Industrial Science & Technology (NMIJ/AIST), Japan Electric Meters Inspection Corporation (JEMIC), and METI-Designated Calibration Laboratories to calibrate the secondary standards operated in the Accredited Calibration Laboratories. The secondary standards are employed to calibrate the reference standards in other Accredited Calibration Laboratories at lower levels within the traceability hierarchy. These calibrated reference standards are then used to calibrate the working standards and measuring instruments applied in precision metrology and various measurement activities. The measurement results of precision metrology can then be made traceable to the national primary standards and the SI. A calibration certificate bearing the JCSS accreditation symbol is provided for calibration results that follow JCSS standards. This certificate validates the traceability of the calibration results to the national primary standards.

　このようなトレーサビリティ体系は日本以外の各国においても確立されており，国際度量衡委員会（CIPM）の国際相互承認取決め（CIPM MRA）によって各国の国家計量標準およびトレーサビリティの同等性を相互に認めるようになっている．国家計量標準間の同等性を確認するために行う比較試験（国際比較）では，同一の校正対象を複数の国家計量標準機関で持ち回り測定するなどして，その結果を比較している．

● 1.3　精密計測用標準器

　精密計測では多くの量が扱われており，それぞれの量が対応する測定標準器が整備されているが，本節では精密計測において特によく使われる時間・周波数，長さおよび温度の測定標準器を取り上げる．Fig.1-3 にそれらの測定標準器をまとめたものを示す．あらゆる計測量の中で，時間と周波数は最も正確に計測できるものである．その計測を行う時計は周期現象を伴う振動子と振動子の振動を数える計数機（カウンタ）からなる．振動子の周波数は振動の時間周期と逆数関係にあり，周波数が決まれば時間も一義的に決まるので，時間計測と周波数計測は同じこととなり，時間標準は周波

Such traceability systems are also established in countries other than Japan. The international equivalence of the traceability systems and the national measurement standards of different countries are recognized through the framework of CIPM MRA (the Mutual Recognition Arrangement of the International Committee for Weights and Measures). The degree of equivalence of the national standards is confirmed by the results of international comparison tests where calibration artifacts are circulated among the national metrology institutes of different countries for the calibration tests.

● 1.3　Measurement standards for precision metrology

Precision metrology encompasses a diverse range of quantities, each of which is governed by a specific set of measurement standards. In this section, the measurement standards of time/frequency, length, and temperature, which are especially important for precision metrology, are presented. Fig.1-3 shows a summary of the measurement standards for these quantities. Among the measurement quantities, time and frequency can be most accurately measured using a clock system comprising an oscillator linked to periodic phenomena and a counter that tallies the oscillation. Given the inverse relationship between the oscillation time period and frequency, the former can be clearly determined after estimating the latter, indicating

数標準とも呼ばれる．日本では，時間・周波数国家標準（UTC(NMIJ) および UTC
(NICT)）は協定世界時（UTC）に同期した水素メーザなどの原子時計によって実現
されており，産業技術総合研究所と情報通信研究機構が運用している．セシウム原子
周波数標準器などの一次周波数標準器は UTC と比較され，UTC を校正している．ま

国家計量標準（特定標準器） Primary standards

協定世界時に同期した原子時計 Atomic clock synchronized to Coordinated Universal Time	協定世界時に同期した原子時計 および光周波数コム装置 Atomic clock and optical frequency comb device synchronized to Coordinated Universal Time	温度定点群実現装置 Temperature fixed points
同期 Synchronization		温度計校正用定点実現装置 Temperature fixed points

特定二次標準器・常用参照標準   Secondary standards/ Reference standards

セシウム原子周波数標準器，ルビジウム原子周波数標準器，水素メーザ，水晶発振器など Cesium frequency standard, Rubidium frequency standard, Hydrogen maser frequency standard, Crystal oscillator, etc.	633 nm ヨウ素分子吸収線波長安定化He-Neレーザ装置，633 nm 実用波長安定化He-Neレーザ装置，校正用ブロックゲージなど 633 nm iodine stabilized He-Ne laser, 633 nm stabilized He-Ne laser, Standard gauge block, etc.	温度計校正用定点実現装置，白金抵抗温度計，放射温度計など Temperature fixed point, Platinum resistance thermometer, Radiation thermometer, etc.

実用標準・一般測定機器 Working standards/ General measuring instruments

原子時計，周波数標準器，信号発生器，発振器，周波数発生器，周波数カウンタ，スペクトラムアナライザ，タイムベースなど Atomic clock, Frequency standard, Signal generator, Oscillator, Frequency generator, Frequency counter, Spectrum analyzer, Time base, etc.	ブロックゲージ，マイクロメータ，座標測定器，形状測定機，レーザ干渉計，変位センサ，角度センサなど Gauge block, Micrometer, Coordinate measuring machine, Surface profiler, Laser interferometer, Displacement sensor, Angle sensor, etc.	白金抵抗温度計，サーミスタ，熱電対，ガラス製温度計，放射温度計など Platinum resistance thermometer, Thermistor, Thermocouple, Glass thermometer, Radiation thermometer, etc.
時間・周波数 Time/ Frequency	長さ Length	温度 Temperature

Fig. 1-3   The measurement standards of time/frequency, length and temperature

their mutual measurability. Similarly, both quantities share identical measurement standards. The primary standards of time and frequency in Japan, the Coordinated Universal Time (UTC)-NMIJ and UTC-National Institute of Information and Communications Technology (NICT), are realized using cesium atomic clocks, such as UTC-synchronized hydrogen masers. They are operated by AIST and NICT. The primary cesium frequency standards are compared with UTC for UTC calibration. Optical clocks exhibiting higher accuracies than cesium frequency standards are also employed to calibrate UTC. Rubidium frequency standards compared with UTC (NMIJ) are operated as the secondary frequency standards, and crystal oscillators are operated as the reference frequency standards in JCSS Accredited Calibration Laboratories. These reference frequency standards are then used to cali-

た，セシウム原子周波数標準器より高精度といわれている光時計も UTC を校正している．JCSS 登録事業者は，UTC(NMIJ) と比較されているルビジウム時計を特定二次標準器として運用し，水晶発振器を常用参照周波数標準器として運用する場合が多い．常用参照標準により，精密計測に用いる信号発生器，周波数カウンタなどが校正されている．

Fig. 1-4 に示すように，長さ l は，光波が伝搬する時間 Δt の計測によって実現される（式(1-1)）．Δt の計測方法は，飛行時間測定法（TOF）による直接計測（式(1-2)，(1-3)）と，干渉信号の位相差 $\Delta\varphi$ を利用する間接計測（式(1-4), (1-5)）に大別される．前者は非常に長い距離の場合に適するが，メートルを具現化する長さ標準器の観

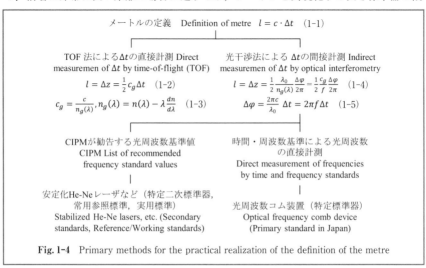

Fig. 1-4 Primary methods for the practical realization of the definition of the metre

brate instruments such as signal generators and frequency counters employed in precision metrology applications.

As shown in Fig. 1-4, the length l is evaluated by measuring the light travel time Δt (Eq. (1-1)). Δt can be directly measured using the time-of-flight (TOF) method (Eqs. (1-2), (1-3)) or can be indirectly measured through the phase difference $\Delta\varphi$ in a length interferometer (Eqs. (1-4), (1-5)). The latter approach is more rational in terms of realizing the metre, while the former only works for very long distances. In the latter approach, precise determination of the optical frequency becomes imperative, achievable through direct measurement by synchronizing the interferometer's light source to the primary frequency standards by an appropriate technique. The primary length standard in Japan (i.e., the optical frequency comb device) is realized based on this approach. Alternatively, the optical frequency value

点から後者の方がより合理的である．この場合，光波の周波数 f を決定する必要があるため，光周波数の時間標準に基づいた測定が必要である．日本の長さに関する国家計量標準（特定標準器）である光周波数コム装置は時間標準を光周波数に変換し，光波（レーザ）周波数の計測値に SI へのトレーサビリティを付与するためのものである．一方，光周波数を直接計測せず，国際度量衡委員会（CIPM）によって勧告されている周波数基準値およびそのレーザ光源を用いて長さを実現する方法もある．多くの国で国家標準として用いられている 633 nm ヨウ素分子吸収線波長安定化 He-Ne レーザ装置はこの方法によるものである．日本ではこのレーザ装置は 2009 年までは特定標準器であり，現在でも光コム（特定標準器）により直接校正される特定二次標準器として運用されている．長さ・幾何学量を中心に扱う精密計測の分野では，特定二次標準器により校正された 633 nm He-Ne レーザが，各種長さ測定機器の校正において基準としての役割をはたしている．

物体の長さが温度の変動に従って変化するので，精密計測では計測時の温度を正確に把握する必要がある．特定標準器として温度定点群実現装置と，特定副標準器としての温度計校正用定点実現装置は，それぞれ産業技術総合研究所と日本電気計器検定

and its associated light source can be selected from the CIPM list of recommended frequency standard values. The 633-nm iodine-stabilized He-Ne laser, which is employed as the primary length standard in most countries, is based on this method; in Japan, it operates as the secondary length standard directly calibrated by the primary standard. The measurement standards based on the 633-nm stabilized He-Ne lasers play an important role in the calibration of length-measuring instruments used in precision metrology.

Temperature measurement is important in precision metrology, as an object's length varies with temperature. The temperature fixed points are operated in AIST as the Primary Standard of Temperature, and in JEMIC as the Sub-Primary Standard in Japan. The temperature fixed points for calibration of contact-type thermometers include the triple point of water, the melting points of indium (In), mercury (Hg), tin (Sn), and zinc (Zn), among others, and those for calibration of radiation thermometers include the meting points of zinc (Zn), aluminum (Al), silver (Ag), and copper (Cu), among others. The SI base unit of thermodynamic temperature is realized using primary thermometers such as the constant volume gas thermometer (CVGT). For practical reasons, the International Temperature Scale (ITS-90) is established by interpolating the temperature fixed points with the sec-

所によって運用されている．接触式温度計校正用温度定点群には水の三重点，インジ
ウム点，水銀点，スズ点および亜鉛点などの温度定点が利用され，また，放射温度計
校正用には銅点，アルミ点，銀点および銅点などの温度定点が利用されている．SIに
おける熱力学温度は定積気体温度計などの一次温度計を用いて実現されるが，実用上
の観点から，これらの温度定点を参照点として，白金抵抗温度計や放射温度計など再
現性のよい二次温度計を用いて補間した国際温度目盛（ITS-90）が温度計の校正に利
用されている．一方，ITS-90と熱力学温度の差を測定する作業も重要となる．

【演習問題】

1-1) SI基本単位キログラムの定義を調べよ．

1-2) 日本の国家計量標準機関の名称を調査せよ．

1-3) 時間・周波数，長さ，温度の国家計量標準で相対不確かさが一番大きいものを
　　 示せ．

1-4) CIPMによって勧告されている未安定化He-Neレーザの周波数基準値とその相
　　 対不確かさを調べよ．

1-5) 熱力学温度を測定するための定積気体温度計で利用されている理想気体の状態
　　 方程式を表せ．

ondary thermometers with better stability, such as platinum resistance and radiation thermometers for thermometer calibration. Notably, measuring the difference
between ITS-90 and the thermodynamic temperature is necessary.

【Problems】

1-1) Investigate the definition of the SI base unit kilogram.

1-2) List the name of the national metrology institute in Japan.

1-3) Among the primary standards of time/frequency, length, and temperature,
which one has the largest relative uncertainty?

1-4) Deduce the CIPM recommended frequency standard value for unstabilized He-
Ne lasers and its relative standard uncertainty.

1-5) Express the general gas equation used in the CVGT to measure the thermodynamic temperature.

第2章　長さスケール

　長さは精密工学において最も重要な物理量の1つである．金属製直尺やノギス，ダイヤルゲージやマイクロメータなど長さを測る機械式スケールが多用されるとともに，高精度な測定にはより高い分解能を有するデジタルスケールが用いられる．本章では，これら長さスケールにおいて重要となる目盛の分解能，および方向弁別の概念について触れた後，各々の長さスケールの構造および測定原理を紹介する．

● 2.1　目盛の分解能と方向弁別

　「分解能」の定義は多岐にわたるが，長さスケールにおいては日本産業規格（JIS Z 8103, ISO/IEC Guide 99 に相当）において「対応する指示値が感知できる変化を生じる，測定される量の最小の変化」と定義されている．すなわち，機械式スケールでは目盛の間隔が分解能に相当する．ノギス，マイクロメータやダイヤルゲージにおいて

Chapter 2　Length Scale

　Length is one of the most important physical quantities in precision engineering. Mechanical length scales such as metal rules, vernier calipers, dial gauges and micrometers are widely used to measure length, and digital scales with higher resolution are used for high-precision measurements. In this chapter, the resolution of the scale and the concept of directional discrimination, which are important in these length scales, are described. The construction and measuring principles of each length scale are also introduced.

● 2.1　Scale resolution and directional discrimination

　Although the definition of 'resolution' could vary widely, for length scales, it is defined in the International Vocabulary of Metrology (VIM) as "smallest change in a quantity being measured that causes a perceptible change in the corresponding indication." In other words, for mechanical scales, the interval between the scale graduations (scale pitch) corresponds to the resolution. In calipers, micrometers

は，後述のとおり目盛をさらに細分化して読み取る工夫により高い分解能が実現され
ている．また，デジタルスケールにおいてはスケール目盛間隔の倍数の周期を有する
読み取り信号をもとに，スケール–読み取り器間の変位を検出する．読み取り信号の波
形は矩形波，三角波もしくは正弦波であることが多く，この信号を Fig. 2-1 のように
電気的に分割（内挿）して変位 x の分解能向上を図る．なおデジタルスケールにおい
ては，1つの読み取り信号だけではスケールの移動方向を判断（方向弁別）できない．
この問題は，Fig. 2-2 に示すように位相が 90°異なる 2 つの読み取り信号（A 相，B
相）を利用すれば解決できる．

● 2.2　機械式スケール

金属製直尺と標準尺

■金属製直尺：直尺は，直線を引くための「定規」と，長さを測る「ものさし」の双方
の機能を有するもので，目盛線間で長さを表す1次元の長さスケールである（Fig. 2-3）.

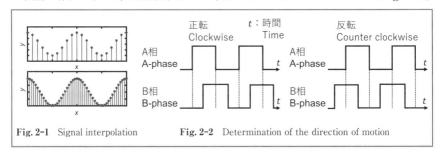

Fig. 2-1　Signal interpolation　　　　**Fig. 2-2**　Determination of the direction of motion

and dial gauges, high resolution is achieved by further subdividing the scale pitch
to be read, as described below in this chapter. In digital scales, the displacement
between the scale and the reading head is detected based on a readout signal with
a period that is a multiple of the scale pitch. The waveform of the readout signal is
often a square wave, a triangular wave or a sine wave, and this signal is electrically
divided (interpolated) as shown in Fig. 2-1 to improve the resolution of displacement x. In digital scales, the direction of motion of the scale cannot be determined
(direction discrimination) from a single readout signal alone; this problem, however,
can be solved by using two readout signals (phase-A and -B) with phase differences of 90° as shown in Fig. 2-2.

● 2.2　Mechanical scales

Metal rules and line scales

■Metal rules: Metal rules are a one-dimensional length scale that has the functions
of both a 'ruler' for drawing straight lines and a 'scale' for measuring length, with

ステンレス，アルミなど金属製のものが多い．目盛端面が基点（目盛の基準）である．
JIS B 7516 により，目盛線の目量は，きりのよい数字（0.5/1/2/5/10 mm など）に規定されており，主な目盛線には基点からの長さが表記される．また，端面の直角度，側面の真直度とそれらの測定方法なども規定されている．

■標準尺：標準尺は，低膨張ガラス基板の表面に，リソグラフィ技術などをもとに等ピッチで目盛を形成した直尺である（Fig. 2-4）．金属製直尺に比べてより精度が高い．基板上には，目盛線とともに標準尺の測定軸を示すアライメントマークが設けられている．線幅は等級ごとに異なるものの高精度なものは数 μm 程度で，その均一性も管理されている．目盛間の長さは，第6章で述べる光干渉計などを用いて標準温度（20℃）下において校正されている．そのため，異なる温度において用いる場合には，標準尺の熱膨張量を算出し補正を加える必要がある．

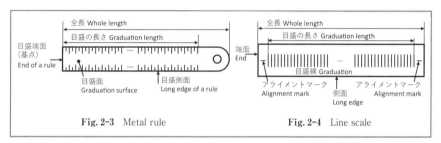

Fig. 2-3 Metal rule   **Fig. 2-4** Line scale

the length expressed between the scale graduations (Fig. 2-3). Many scales are made of metal such as stainless steel or aluminum. The scale pitch is specified by JIS as 0.5, 1, 2, 5 or 10 mm, and the length from the end of a rule is indicated on the main graduation lines. The perpendicularity of end faces, straightness of side faces and their measurement methods are also specified by JIS.

■Line scales: Line scales are made of a low-expansion glass substrate on which scale graduations are formed at an equal pitch based on lithographic techniques (Fig. 2-4). It is more accurate than metal rules. An alignment mark is provided on the substrate to indicate the measurement axis of the line scale together with the scale graduations. Although the line width differs from grade to grade, high-precision scales have a width of only a few micrometres. The uniformity of the line width is also controlled. The lengths between the scale graduations are calibrated at a standard temperature (20℃) using a laser interferometer (described in Chapter 6) or equivalent equipment. Therefore, when used at different temperatures, it is necessary to calculate the thermal expansion of the line scale and make corrections.

ノギス　ノギスは一般的な加工現場において多用される機械式長さスケールである．その測定対象に応じて様々なバリエーションがあるが，一般に用いられるノギスはFig. 2-5のように，外側ジョウおよび内側ジョウからなる．右側のジョウはスライダと呼ばれ，測定軸方向に可動である．本尺とともに副尺（バーニヤスケール）を備えており，これにより本尺の目盛間隔を超える分解能での長さ測定が可能である（Fig. 2-6）．本尺には長さLを$kN-1$分割して得られる間隔S_Mの目盛を配置し，副尺には長さLをN分割して得られる間隔S_Vの目盛を配置する．このとき，式(2-1)および(2-2)が得られ，これらの式から式(2-3)が導出される．i番目の本尺目盛と副尺目盛間の間隔ΔS_{ik}は，Fig. 2-6より，式(2-4)のとおりとなる．この式中のΔsが，副尺を用いた長さ測定の分解能となる．

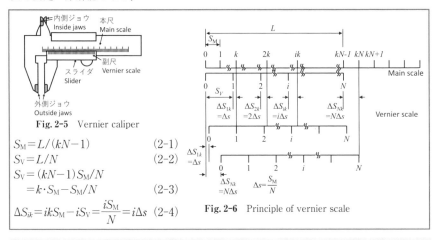

Fig. 2-5　Vernier caliper

$$S_M = L/(kN-1) \qquad (2\text{-}1)$$
$$S_V = L/N \qquad (2\text{-}2)$$
$$S_V = (kN-1)S_M/N$$
$$\quad = k \cdot S_M - S_M/N \qquad (2\text{-}3)$$
$$\Delta S_{ik} = ikS_M - iS_V = \frac{iS_M}{N} = i\Delta s \quad (2\text{-}4)$$

Fig. 2-6　Principle of vernier scale

Vernier caliper　Vernier calipers are mechanical length scales that are widely used in general machine shops. Although there are various variations depending on the object to be measured, commonly used vernier calipers consist of outside jaws and inside jaws, as shown in Fig. 2-5. The right-hand jaw is referred to as the slider and is movable in the direction of the measuring axis. Together with the main scale, it is equipped with a vernier scale, which enables length measurement with a resolution exceeding the pitch of the main scale (Fig. 2-6). The main scale has a scale with a pitch S_M obtained by dividing the length L into $kN-1$, and the vernier scale has a scale with a pitch S_V obtained by dividing the length L into N. Eqs. (2-1) and (2-2) can thus be obtained and Eq. (2-3) can be derived from these equations. From Fig. 2-6, the interval ΔS_{ik} between the ith main scale and the vernier scale can be obtained as Eq. (2-4), where Δs is the resolution of the length measurement using the vernier caliper.

ダイヤルゲージ　ダイヤルゲージは，てこ機構，または歯車列をもとに測定子の直線変位をダイヤルの回転角度変位に拡大して高精度な長さ測定を実現する機構を有する．プランジャー式，てこ式に大別される．いずれの方式でもバネ機構が採用されている．その構造上，測定子が押し込まれた状態で測定がなされることから，測定子がダイヤルゲージの測定レンジを超えて動くことのないよう留意する必要がある．ブロックゲージなどの長さ標準と組み合わせて用いることで対象物の厚さ測定に適用でき，高精度な走査機構/回転機構との併用で，真直度/回転偏心測定にも適用できる．

■プランジャー式ダイヤルゲージ：プランジャー式ダイヤルゲージでは，歯車列をもとに測定子の直線変位をダイヤルの回転角度変位に拡大，表示する（Fig. 2-7）．測定子はピッチ P のラック G_1 と同軸で取り付けられており，このラックとピニオン G_2

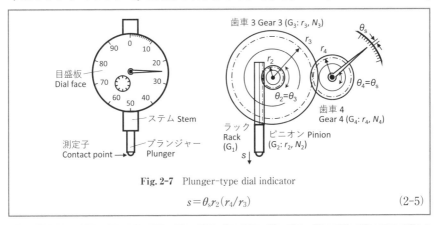

Fig. 2-7　Plunger-type dial indicator

$$s = \theta_s r_2 (r_4 / r_3) \tag{2-5}$$

Dial gauge　Dial gauges have a mechanism that magnifies the linear displacement of the measuring object to a rotational angular displacement of the dial based on a gear train or a lever mechanism to realize highly accurate length measurement. They are broadly classified into plunger type and lever type. Both types of dial gauges employ a spring mechanism. Due to their structures, the measuring element is pushed in to make measurements, so attention should be paid to ensure that the measuring object does not move beyond the measuring range of the dial gauge. Dial gauges can be used in combination with length standards such as gauge blocks to measure the thickness of objects. They can also be used in combination with a high-precision scanning or rotating mechanism to measure straightness or eccentricity.

■Plunger-type dial gauge: In a plunger-type dial gauge, the linear displacement of the measuring object is magnified and displayed as a rotational angular displacement of the dial based on the gear train (Fig. 2-7). The measuring object is mount-

（ピッチ円半径 r_2, 歯数 N_2）が噛み合っている．ピニオンは回転軸を同一とする歯車 $G_3(r_3, N_3)$ に取り付けられており，G_3 と噛み合う歯車 $G_4(r_4, N_4)$ の回転中心にダイヤルゲージのポインタが取り付けられている．ダイヤルゲージの文字盤上における角度スケールのピッチを θ_S とすると，これに相当する歯車 4 の回転角 θ_4 は $\theta_4=\theta_S$ となる．歯車 4 と歯車 3 は噛み合っているため，歯車 3 の回転角 θ_3 は $r_4\theta_4=r_3\theta_3$ を満たす．なお，G_2 は G_3 と回転軸を同一として G_3 に固定されていることから，その回転角 θ_2 は θ_3 と一致する．回転角 θ_2 に相当する測定子の直線変位 s は，$s=r_2\theta_2$ である．したがって，s と θ_S との間には式（2-5）が成り立つ．このように，s および θ_S を設定すると，幾何的条件をもとに各歯車のパラメータを定めることができる．

■てこ式ダイヤルゲージ（テストインジケータ）：てこ式ダイヤルゲージでは，てこ機構をもとに測定子先端の変位を拡大した後，歯車列をもとにその変位をさらに拡大表

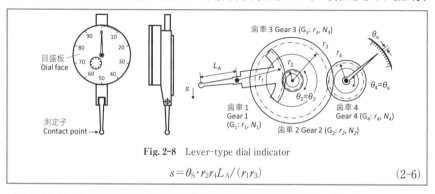

Fig. 2-8 Lever-type dial indicator

$$s=\theta_S \cdot r_2 r_4 L_A/(r_1 r_3) \qquad (2\text{-}6)$$

ed so that it becomes coaxial with the rack (G_1) of pitch P, which meshes with the pinion $(G_2$, pitch circle radius: r_2, number of teeth: $N_2)$. The pinion is mounted on gear $G_3(r_3, N_3)$ having the same axis of rotation, and the pointer of the dial gauge is mounted at the centre of rotation of gear $G_4(r_4, N_4)$ meshing with G_3. If the pitch of the angle scale on the dial of the dial gauge is denoted by θ_S, the angle of rotation θ_4 of G_4 corresponding to this pitch is the same as θ_S. Since G_3 and G_4 are meshing with each other, the rotational angle θ_3 of G_3 satisfies $r_4\theta_4=r_3\theta_3$. It should be noted that G_2 is fixed to G_3 with the same axis of rotation. Therefore, its angle of rotation θ_2 is equal to θ_3. The linear displacement s of the measuring object corresponding to the rotational angle θ_2 is $s=r_2\theta_2$. Therefore, Eq. (2-5) holds between s and θ_S. As a result, once s and θ_S are fixed, the parameters of each gear can be determined based on the geometrical conditions.

■Lever-type dial gauge（test indicator）：In a lever-type dial gauge, the displacement of the tip of the measuring object is magnified based on the lever mechanism,

示する（Fig. 2-8）．測定子の回転中心から測定子先端までの距離をL_A，回転中心から歯車G_1のピッチ円までの距離をr_1とし，これに噛み合う歯車G_2が回転軸を同一とする歯車G_3に取り付けられている．また，G_3と噛み合う歯車G_4の回転中心にダイヤルゲージのポインタが取り付けられている．プランジャー式ダイヤルゲージでの計算と同様，幾何的関係より測定子の直線変位sと文字盤上における角度スケールのピッチθ_Sとの間の関係式(2-6)を得ることができ，sおよびθ_Sを設定することで各歯車のパラメータを定めることができる．なお，測定子に角度をつけた設計も可能であるが，測定子の感度方向と変位の方向は$10°$以内に設定するのが望ましい．

マイクロメータ　マイクロメータは，高精度ネジの等ピッチ性を利用した長さスケールである（Fig. 2-9）．主に，フレーム，アンビル，スピンドル，スリーブ，シンブルおよびラチェットから構成されており，アンビルとスピンドルの間に測定対象を挟み込

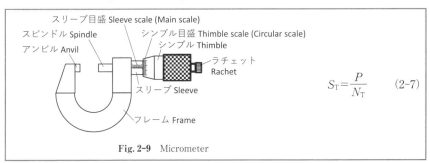

スリーブ目盛 Sleeve scale (Main scale)
スピンドル Spindle
アンビル Anvil
シンブル目盛 Thimble scale (Circular scale)
シンブル Thimble
ラチェット Rachet
スリーブ Sleeve
フレーム Frame

$$S_T = \frac{P}{N_T} \qquad (2\text{-}7)$$

Fig. 2-9　Micrometer

and then the displacement is further magnified based on the gear train (Fig. 2-8). Now, denote the distance from the centre of rotation of the measuring object to the tip of the measuring object as L_A, and the distance from the centre of rotation to the pitch circle of gear G_1 as r_1. gear G_2 meshing with G_1 is attached to gear G_3 with the same rotational axis. The pointer of the dial gauge is attached to the centre of rotation of the gear G_4 meshing with G_3. In the same manner as the calculations of the plunger-type dial gauge, the relationship between the linear displacement s of the measuring element and the pitch θ_S of the angle scale on the dial can be obtained as Eq. (2-6) from the geometric relationship, and the parameters of each gear can be determined by fixing s and θ_S. Although an angled stylus design is possible, the direction of sensitivity and displacement of the stylus is expected to be set within $10°$.

Micrometer　A micrometer is a length scale that utilizes the equal pitch of the threads of a high-precision screw (Fig. 2-9). It mainly consists of a frame, anvil, spindle, sleeve, thimble, and ratchet. The length is measured by placing the object

んで長さを測定する．スリーブ円筒面にメインスケールが設けられており，その目盛
S_M はマイクロメータ内のネジのピッチ P によって決まる．またシンブル円筒面には
シンブルスケールが設けられており，その目盛間隔 θ_T はシンブルスケール1周（360°）
の分割数 N_T により決まり $\theta_T = 360°/N_T$ となる．また，マイクロメータの目盛分解能
S_T は N_T によって式(2-7)のように決まる．バーニヤスケールとの併用でさらに高い分
解能が得られる．なお，マイクロメータはアッベの原理を満たしている（第11章参照）．

● 2.3　デジタルスケール

デジタルスケールでは，スケールと読み取りヘッドの間の相対変位が検出される．
アプリケーションごとに各々の配置は異なるものの，一般には一方をシステム内で固
定した状態で用いられる．現在市販されているデジタルスケールは，スケール上に等
ピッチで設けられたパターンの読み取り方式により電磁誘導式，磁気式あるいは光学
式に大別される．

電磁誘導式スケール　電磁誘導式スケールでは，スケール上にピッチ P のスケールコ
イル列が形成されている（Fig. 2-10(a)）．スケールコイルはいずれも並列接続され，時

to be measured between the anvil and the spindle. The main scale is provided on
the cylindrical surface of the sleeve and its scale pitch S_M is determined by the pitch
P of the screw in the micrometer. There is also a thimble scale on the thimble cylindrical surface, whose scale pitch θ_T is $\theta_T = 360°/N_T$. The resolution S_T of the micrometer is determined by the number of divisions N_T of one round (360°) of the
dimple scale, as shown in Eq. (2-7). A higher resolution can be obtained by combining with a vernier scale. A micrometer satisfies Abbe's principle (see Chapter 11).

● 2.3　Digital scales

In digital scales, the relative displacement between the scale and the reading
head is detected. Although the arrangement of each differs from application to application, generally, one of them is fixed in the system. The digital scales currently
available on the market are broadly classified as inductive, magnetic, or optical type,
depending on the reading method of the scale graduations.

Electromagnetic induction scale　In an electromagnetic induction scale, a row of
scale coils with a pitch P is formed on the scale (Fig. 2-10(a)). Assuming that the

間 t で変動する電流 $I_0(t)$ が流れているものとすると，各コイルで発生する磁場 $B(t)$ は透磁率を μ_0 として式(2-8)で与えられる．読み取りヘッドには検出コイルが構成されており，スケールコイル列上を通過する際に電磁誘導によりその両端に誘導電圧 $\varepsilon(t)$ が発生する．検出コイルの長さと幅を $P/2$ とし，スケールコイルが重なり合う領域の幅を Δx とすると，アンペアの周回積分の法則およびファラデーの電磁誘導の法則に基づき，$\varepsilon(t)$ は式(2-9)で与えられる．検出コイルとスケールコイル列の相対変位に応じて変化する誘導電圧（Fig. 2-10(b)）を得ることで，Δx が検出できる．

磁気式スケール　磁気式スケールでは，磁性媒体上に間隔 P で設けた磁極列パターンをスケールとして用い，これを磁気抵抗が磁界によって変化する磁気抵抗効果を利用したホール素子あるいは MR 素子で読み取ることで，水平方向の相対変位を検出する．

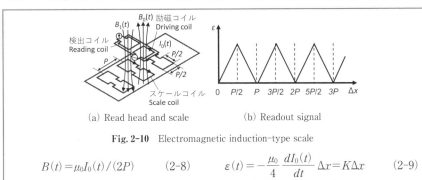

(a) Read head and scale　　　　(b) Readout signal

Fig. 2-10　Electromagnetic induction-type scale

$$B(t) = \mu_0 I_0(t)/(2P) \qquad (2\text{-}8)$$

$$\varepsilon(t) = -\frac{\mu_0}{4}\frac{dI_0(t)}{dt}\Delta x = K\Delta x \qquad (2\text{-}9)$$

scale coils are all connected in parallel and that a current $I_0(t)$, which varies with time t, flows through them, the magnetic field $B(t)$ generated by each coil is given by Eq. (2-8), where μ_0 is magnetic permeability. The reading head consists of reading coils, which generate an induced voltage $\varepsilon(t)$ at both ends due to electromagnetic induction as they pass over the scale coils. When the length and width of the reading coil are denoted by $P/2$, and the width of the area where the scale coils overlap with each other is denoted by Δx, $\varepsilon(t)$ is given by Eq. (2-9) based on Ampère's circuital law and Faraday's law of induction. Δx can thus be detected by obtaining an induced voltage that varies as shown in Fig. 2-10(b) according to the relative displacement Δx.

Magnetic scale　The magnetic scale detects relative displacement in the horizontal direction by using a magnetic pole pattern with an interval P on a magnetic medium as a scale, which is read by a Hall sensor or MR sensor using the magneto-resistive effect, in which the magnetic resistance changes with the magnetic field. By

半ピッチ分だけ位相をずらして MR 素子を複数配置することで，A/B 相信号を得て方向弁別を実現する．

光学式スケール（スリット方式，モアレ式）　スリット方式の光学式スケール（Fig. 2-11 (a)）は，光源，透過型スケールおよび受光素子から構成されており，光源および受光素子は一体としてスケールに対して相対運動する．一定のピッチ P で線状パターンが刻まれたスケールが移動した際の，光源からの光の通過/遮蔽を受光素子で検出することで相対変位 Δx を検出する．モアレ式の光学式スケール（Fig. 2-11 (b)）では，さらにレチクル格子を用い，レチクル格子およびスケールに刻まれたスリットをもとに，受光素子上にモアレを発生させる．受光素子を複数配置し，レチクル格子のパターンを工夫して位相が 90°ずれた読み取り信号を得ることで，方向弁別も可能となる．

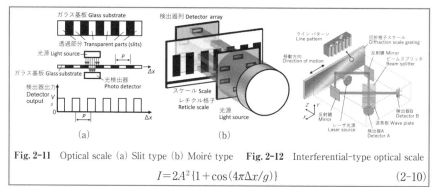

Fig. 2-11　Optical scale (a) Slit type (b) Moiré type　**Fig. 2-12**　Interferential-type optical scale

$$I = 2A^2\{1 + \cos(4\pi\Delta x/g)\} \qquad (2\text{-}10)$$

arranging multiple MR elements with their phases shifted by half a pitch, A/B phase signals are obtained, and direction discrimination is achieved.

Optical scale (slit type, moiré type)　The slit-type optical scale (Fig. 2-11(a)) consists of a light source, a transmission type scale and a photodetector. The light source and light-receiving element move relative to the scale. The relative displacement Δx is detected by detecting the passage or shielding of light from the light source by the photodetector when the scale with linear patterns having a pitch P moves. In a moiré-type optical scale (Fig. 2-11(b)), a reticle grating is employed to generate moiré on the photodetector based on the slits prepared on the reticle grating and scale. Directional discrimination is also possible by arranging multiple photodetectors and designing the pattern of the reticle grating to obtain a reading signal with a phase shift of 90°.

光干渉式リニアスケール　回折光干渉型の光学式リニアエンコーダ（Fig. 2-12）では，μm～サブ μm 級のラインパターンを有する短ピッチ長さスケールが用いられる．ラインパターンの間隔（ピッチ）が g であるスケールに対し，波長 λ の測定光を入射した際に得られる ±1 次回折光を重ね合わせて生成した干渉信号（光強度：I）を読み取り信号として利用することで，高い分解能での変位検出が実現する．反射回折光の複素振幅を A とすると，スケール格子の変位 Δx に伴い式(2-10)のように光強度 I が正弦波状に変化する干渉信号が得られる．なお，偏光光学素子を利用して位相差 $\pi/2$ rad を有する2つの干渉信号を得ることで，方向弁別も可能となる．この干渉信号に対して電気的な内挿処理を行うことで，サブ nm 超の高分解能が実現されている．

光干渉式サーフェスエンコーダ　光干渉式リニアスケールのスケールを2軸スケール

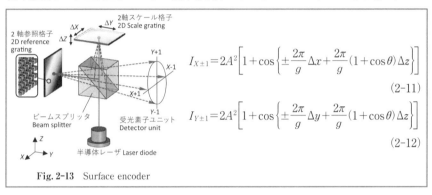

$$I_{X\pm1}=2A^2\left[1+\cos\left\{\pm\frac{2\pi}{g}\Delta x+\frac{2\pi}{g}(1+\cos\theta)\Delta z\right\}\right] \tag{2-11}$$

$$I_{Y\pm1}=2A^2\left[1+\cos\left\{\pm\frac{2\pi}{g}\Delta y+\frac{2\pi}{g}(1+\cos\theta)\Delta z\right\}\right] \tag{2-12}$$

Fig. 2-13　Surface encoder

Interferential scanning-type optical linear scale　In an interferential scanning-type optical linear encoder (Fig. 2-12), a scale grating having scale graduations with a pitch of μm to sub-μm is employed. Displacement measurement with high resolution is achieved by using the interference signal (light intensity: I) generated by superimposing the positive and negative first-order diffracted beams obtained when a measurement light of wavelength λ is made incident on a scale with a pitch of g as the readout signal. The light intensity I is given by Eq. (2-10), where A is the complex amplitude. An interferential signal that varies sinusoidally with the displacement Δx of the scale grating is obtained. Direction discrimination is also possible by using polarization optical components to obtain two interference signals with a phase difference of $\pi/2$ rad. High resolution of sub-nanometre or better can be achieved by electrical interpolation of the interference signals.

Optical interferential-type surface encoder　When the scale of an optical interference linear scale is changed to a two-dimensional (2D) diffraction scale grating, a

格子に変更すると，1本の測定レーザ光で XY 面内2自由度の変位を同時に検出できるサーフェスエンコーダが実現する（Fig.2-13）．2軸スケール格子で発生した ±1 次回折光と，光学ヘッド内に新たに設けた2軸参照格子で発生した ±1 次回折光とを重畳すると，式(2-11)および(2-12)に示す干渉信号が得られる．これらの式をもとに演算することで，面内変位 Δx，Δy および面外変位 Δz を同時に検出可能となる．

【演習問題】

2-1) ノギスにおいて，目盛ピッチ S_M＝1 mm を有する本尺を用いた場合に分解能 Δs ＝0.05 mm を得るための副尺の諸元を求めよ．

2-2) Fig.2-7 に示すプランジャー式ダイヤルゲージにおいて，分解能 s＝10 μm を得るための各歯車およびダイヤルゲージ文字盤のパラメータを定めよ．

2-3) マイクロメータにおいて，ディンプルスケール1周（360°）の分割数 N_T＝50，マイクロメータ内のネジのピッチ P＝0.5 mm としたときの分解能 S_T を求めよ．

2-4) 格子ピッチ g＝1 μm の回折スケール格子を有する光干渉式リニアスケールにおいて，分解能 1 nm を実現するために必要となる干渉信号の内挿分割数を求めよ．

surface encoder that can simultaneously detect two-dimensional in-plane displacements with a single measurement laser beam can be realized (Fig. 2-13). The interference signals shown in Eqs. (2-11) to (2-12) are obtained by superimposing the positive and negative first-order diffracted beams generated by the 2D scale grating and those generated by the newly installed 2D reference grating in the optical head. By performing arithmetic operations based on these equations, the in-plane displacements Δx and Δy, as well as the out-of-plane displacement Δz, can be detected simultaneously.

【Problems】

2-1) Find the specifications of the vernier scale of a vernier caliper to obtain a resolution of Δs＝0.05 mm when using a main scale with a scale pitch of S_M＝1 mm.

2-2) For the plunger-type dial gauge shown in Fig. 2-7, determine the parameters of each gear and dial gauge dial to obtain a resolution of s＝10 μm.

2-3) In a micrometer, find the resolution S_T when the number of divisions N_T 50 for one revolution (360°) of the dimple scale and the pitch of the screw in the micrometer P＝0.5 mm.

2-4) Find the number of interpolation divisions of the interference signal required to achieve a resolution of 1 nm on an optical interferometric linear scale with a diffraction scale grating of grating pitch g＝1 μm.

第3章　角度スケール

　角度は長さとともに精密工学において最も重要な物理量の1つである．この角度の測定に用いられる角度スケールには，1周360°を高精度に分割して得られる円周分割法に立脚したもの，長さ測定を組み合わせた正弦演算により得られる正弦法則に立脚したものが挙げられる．本章では，角度の定義，工業現場において用いられる角度標準について触れた後，各々の角度スケールの原理および構造を紹介する．

● 3.1　角度の定義と角度標準

　SI単位系において，角度は単位radを用いて表現される．組立単位として便宜上は位置づけられているが，実質的には無次元量である．角度は物理量によらず，1周360°（2π rad）と幾何学的に定義されている．日本では，産業技術総合研究所が所有する角度測定装置が特定標準器としてトレーサビリティ体系の頂点に位置し，校正事業者が

Chapter 3　Angle Scale

　Angle, along with length, is one of the most important physical quantities in precision engineering. Angle scales used to measure angles include those based on the self-proving principle of dividing the circle, which is obtained by precisely dividing a 360° circumference, and those based on the sine principle, which is obtained by trigonometric operation combined with length measurement. In this chapter, the definition of angle, angle standards used in industry, and the principle and structure of each angle scale are introduced.

● 3.1　Definition of angle and angle standard

　In the SI system of units, angles are expressed using the unit rad. Although it is positioned as a derived unit for convenience, it is practically a dimensionless quantity. Angles are defined geometrically as 360° (2π rad) around a circle, independent of physical quantities. In Japan, the angle measuring device owned by the National Institute of Advanced Industrial Science and Technology (AIST) is located at the

有するロータリエンコーダおよびロータリエンコーダ自己校正装置をもとにユーザが
用いる角度スケールが校正される.

　なお,一般の工業現場において利用可能な角度の標準器としては,より上位の測定
器で校正されたロータリエンコーダが用いられることが多い.より簡便な手法として,
長さ測定におけるブロックゲージのように,決まった傾斜角を有する角度ゲージ
(Fig. 3-1(a)) も存在するが,直角 (90°) など特殊な角度を除き,その入手は決して
容易ではない.そのため一般には,2点間の長さ L が高精度に保証されたサインバー
をブロックゲージと組み合わせることで任意の角度標準を作り出すことが多い
(Fig. 3-1(b)).平面度が保証された定盤上にサインバーを配置するとともに,その片
足に挟むブロックゲージの高さ h を調整することで,式(3-1)をもとに任意の傾斜角 θ
を作り出すことができる.

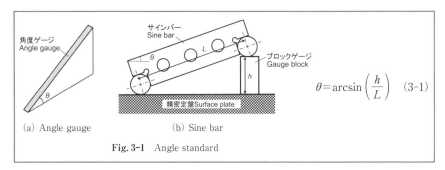

角度ゲージ
Angle gauge

サインバー
Sine bar

ブロックゲージ
Gauge block

精密定盤 Surface plate

$$\theta = \arcsin\left(\frac{h}{L}\right) \quad (3\text{-}1)$$

(a) Angle gauge　　　　(b) Sine bar

Fig. 3-1　Angle standard

top of the traceability system as a specified standard, and the angle scale used by
users is calibrated based on the rotary encoder and rotary encoder-based self-calibration device owned by the calibration laboratories.

　In general, rotary encoders calibrated by a higher-level measuring instrument
are often used as angle standards that can be used at industrial sites. As a simpler
method, angle gauges with a fixed inclination angle (Fig. 3-1(a)), like gauge blocks
in length measurement, also exist. However, they are not easily available except for
special angles such as a right angle (90°). Therefore, in general, a sine bar with a
guaranteed length between two points with high accuracy is often combined with
gauge blocks to create an arbitrary angle reference (Fig. 3-1(b)). A sine bar has a
guaranteed length of L between two points with high accuracy. By placing the sine
bar on the flatness-guaranteed surface plate and adjusting the height h of the
gauge blocks placed beneath one of its legs, an arbitrary inclination angle θ can be
created based on Eq. (3-1).

● 3.2　円周分割法

円周分割法では，幾何学的に定義される1周360°という性質をもとに，円周を高精度に分割することで角度スケールが得られる．ここでは，円周を n 分割する角度スケールの校正を考える．Fig. 3-2のように，目盛1～n の n 個の目盛を有する円周スケールに対して2つの検出器 A，B を開き角 α（おおよそ $360°/n$）で配置する．いま，円周スケールの各目盛間隔をそれぞれ θ_1～θ_n とし，まずは検出器 A に対して目盛1を合わせ，このときの検出器 B と目盛2とのずれ量 δ_1 を式(3-2)のように求める．次に，円周スケールを回転させ，検出器 A に対して目盛2を合わせ，このときの検出器 B と目盛3とのずれ量 δ_2 を式(3-3)のように求める．円周スケールを回転しながらこの作業を繰り返し，検出器 A に対して目盛 i（$i=1$～$n-1$）を合わせ，このときの検出器 B と目盛 $(i+1)$ とのずれ量 δ_i を式(3-4)のように求める．最後に検出器 A に対して目盛 n

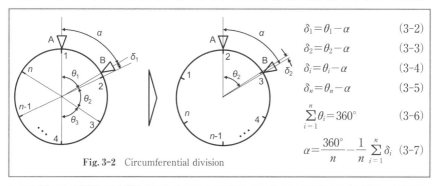

$$\delta_1 = \theta_1 - \alpha \qquad (3\text{-}2)$$

$$\delta_2 = \theta_2 - \alpha \qquad (3\text{-}3)$$

$$\delta_i = \theta_i - \alpha \qquad (3\text{-}4)$$

$$\delta_n = \theta_n - \alpha \qquad (3\text{-}5)$$

$$\sum_{i=1}^{n} \theta_i = 360° \qquad (3\text{-}6)$$

$$\alpha = \frac{360°}{n} - \frac{1}{n}\sum_{i=1}^{n} \delta_i \quad (3\text{-}7)$$

Fig. 3-2　Circumferential division

● 3.2　The self-proving principle of dividing the circle

This principle is based on the fact that a circle can be divided into any number of parts and the divisions are self-proved upon closing of the circle. Now we consider the calibration of an angle scale that divides the circumference into n segments. As shown in Fig. 3-2, two detectors A and B are placed at an opening angle α (approximately $360°/n$) with respect to a circumferential scale with n graduations from 1 to n. The interval between each pair of neighboring scale graduations is θ_1 to θ_n, respectively. At first, the first graduation is aligned with detector A, and the deviation δ_1 between detector B and the second graduation is calculated as Eq. (3-2). Secondly, rotate the circumferential scale, align the second graduation with respect to detector A, and determine the deviation δ_2 between detector B and the third scale as Eq. (3-3). Repeat this process while rotating the circumferential scale, align the ith graduation ($i=1$ to $n-1$) with respect to detector A, and determine the amount of deviation δ_i between detector B and the $(i+1)$th graduation as Eq. (3-4). Finally, the nth scale graduation is aligned with respect to detector A, and the devi-

を合わせ，このときの検出器Bと目盛1とのずれ量 δ_n を式(3-5)のように求める．なお，1周で 360° という幾何的条件から，式(3-6)が得られる．以上の方程式を解くことで，検出器 A〜B 間の開き角 α が式(3-7)のように求まる．この値を個々の式に代入することで，目盛間隔 θ_1〜θ_n が得られる．この原理をもとに円周を逐次分割することで，高精度かつ分解能の高い目盛を有する角度スケールが得られる．円周分割法に立脚した角度センサとしては，ロータリエンコーダが挙げられる．

ロータリエンコーダ（磁気式，光学式）　ロータリエンコーダは，第2章で取り扱ったデジタルスケールの直線スケールを円盤/円周スケールにしたもので (Fig. 3-3)，リニアスケールと同様，相対角度変位を検出するインクリメンタル型と，絶対角度を検出するアブソリュート型がある．

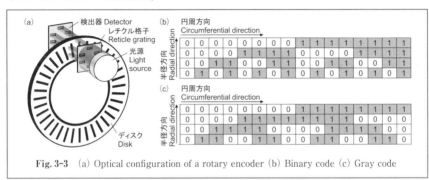

Fig. 3-3　(a) Optical configuration of a rotary encoder (b) Binary code (c) Gray code

ation δ_n between detector B and the first graduation is calculated as Eq. (3-5). The geometric condition of 360° per revolution yields Eq. (3-6). By solving the above equations, the opening angle α between detectors A and B can be obtained as shown in Eq. (3-7). By substituting this value into the individual equations, the scale interval θ_1 to θ_n can be obtained. Based on this principle, angle scales with high accuracy and resolution can be obtained by dividing the circumference of the circle sequentially. A rotary encoder is an example of an angle sensor based on the division of the circle.

Rotary encoder (magnetic, optical)　A rotary encoder has a disk/circumferential scale that is similar to a digital linear scale treated in Chapter 2 (Fig. 3-3). In the same manner, as linear scales, there are two types of rotary encoders: an incremental type that detects relative angular displacement and an absolute type that detects absolute angle position.

■インクリメンタル型：円周方向に同角度ピッチで並べた目盛を有する円盤／円周スケールが用いられる．検出手法は第 2 章で述べたデジタルスケールと同様であり，磁気式，電磁誘導式および光学式のものがある．スリット方式のものはシンプルな装置構成で実現できるが分解能が制限される．一方，光干渉式のロータリエンコーダでは，読み取りヘッドの装置構成が複雑になるが，高分解能化が期待できる．

■アブソリュート型：1 周 360° をビット数で分割した値が検出分解能となる．従来はビット数分のトラックを並列に並べたバイナリスケール（Fig. 3-3(b)）またはグレイスケール（Fig. 3-3(c)）が用いられていたが，トラック数の制約があることから実現可能な分解能に限界があった．一方，M コードなど，トラック数によらず高いビット数での分割を実現できる手法が開発されており，そのコード読み取り用画像センサの高速化・高分解能化とも相まって，高帯域・高精度な絶対角度測定が実現されている．アブソリュート型のロータリエンコーダでは，絶対角度位置が検出可能である．そのためインクリメンタル方式においてアプリケーションでの利用時に必須となる初期化動作を行う必要がなく，また装置電源の復帰時に絶対角度位置を瞬時に把握してオペ

■Incremental type：A disk/circumferential scale with graduations arranged at the same angular pitch in the circumferential direction is used. The detection method is the same as that of the digital length scales described in Chapter 2; magnetic, electromagnetic induction and optical methods are available. A slit-type rotary encoder can be realized with a simple device configuration, although its resolution is limited. On the contrary, an optical interferometric rotary encoder can achieve higher resolution, although it requires a more complex read head configuration.

■Absolute type：In an absolute-type rotary encoder, the resolution will be the value obtained by dividing 360° per revolution by the number of bits. Conventionally, binary scales (Fig. 3-3(b)) or gray scales (Fig. 3-3(c)) with parallel tracks of the same number of bits have been used, although the number of tracks is limited; this restricted the feasible resolution of angle measurement. On the other hand, M-code and other methods have been developed that can achieve a high bit number of divisions regardless of the number of tracks, and together with the increased speed and resolution of image sensors used to read these codes, high bandwidth and high precision absolute angle measurement has been realized. Since absolute rotary encoders can detect absolute angular position, an initialization operation, which is necessary for the incremental type, will not be required. In addition, the absolute angle position can be instantly determined when the equipment power is turned back on to start operation; this contributes to improving the throughput and safety of the

レーションを開始できるなど，装置のスループットと安全性の向上に寄与することか
ら，その重要性が高まっている．

● 3.3　正弦法則

正弦法則は，直線変位センサで得られた変位情報をもとに，幾何的条件からなる演
算を行うことで測定対象の微小角度変位 $\Delta\theta$ を得る方法である．大別すると，一対の
変位計を用いて測定対象面の異なる2点で測定した直線変位 Δy をもとに $\Delta\theta$ を得る方
法（Fig. 3-4）と，測定面からの反射光の反射角変動として，それを捕捉する受光素子
上での反射光の変位 Δx に換算して $\Delta\theta$ を検出する方法（Fig. 3-5）が挙げられる．

一対の変位計で検出する方法　距離を置いて配置した複数の長さスケールで読み取っ
た変位から算出する手法である．いま，Fig. 3-4 のように，測定対象の変位を，既知

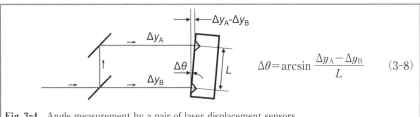

$$\Delta\theta = \arcsin \frac{\Delta y_A - \Delta y_B}{L} \qquad (3\text{-}8)$$

Fig. 3-4　Angle measurement by a pair of laser displacement sensors

equipment. For these reasons, the importance of absolute-type rotary encoders is
increasing.

● 3.3　The sine principle

The sine principle obtains a small angular displacement $\Delta\theta$ of a measurement target by performing trigonometric operations based on geometric conditions, using
displacement information obtained with single or multiple displacement sensors.
There are major two configurations; in the first configuration, $\Delta\theta$ is obtained based
on the linear displacement Δy measured at two different points on the surface to be
measured using a pair of displacement transducers, while in the second configura-
tion $\Delta\theta$ is detected as the variation of position Δx of reflected light from the surface
to be measured on a detector.

A method employing a pair of displacement sensors　This method calculates the
angular displacement from the readings of multiple length scales placed at a distance from each other. Now we consider the case where the angular displacement

の間隔 L で配置した2つのレーザ干渉変位計 A, B で検出する場合を考える．測定対象の微小角度変位 $\Delta\theta$ 発生前後のレーザ干渉変位計の読みの変化量をそれぞれ Δy_A, Δy_B とすると，$\Delta\theta$ は式(3-8)で算出できる．なお，本手法で高精度な微小角度変位検出を実現するためには，予め2つのレーザの間隔 L を高精度に求めておく必要がある．

受光素子上での反射光の変位に換算して検出する方法　Fig.3-5 のように測定対象に微小角度変位 $\Delta\theta$ が発生すると，反射光の反射角は $2\Delta\theta$ となり，距離 L だけ離れた観測面における反射光の直線変位 Δx は式(3-9) のとおりとなる．したがって，観測面にイメージセンサを配置し，Δx を検出することで $\Delta\theta$ が得られる．本手法は光てこ法とも呼ばれる．なおこの場合，Δx は測定対象と観測面との間の距離 L によっても変

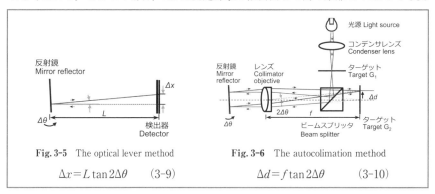

Fig. 3-5　The optical lever method

$$\Delta x = L \tan 2\Delta\theta \qquad (3\text{-}9)$$

Fig. 3-6　The autocolimation method

$$\Delta d = f \tan 2\Delta\theta \qquad (3\text{-}10)$$

of an object to be measured is detected by two laser interferometers A and B placed at a known distance L, as shown in Fig. 3-4. When the change in the readings of the laser interferometers before and after the occurrence of a small angular displacement $\Delta\theta$ of the measurement target is expressed by Δy_A and Δy_B, respectively, $\Delta\theta$ can be calculated by using Eq. (3-8). In order to achieve highly accurate detection of small angular displacement with this method, the distance L between two length scales should be determined precisely in advance.

A method by converting the angular displacement into the displacement of reflected light on a detector　When a small angular displacement $\Delta\theta$ occurs on the measurement target as shown in Fig. 3-5, the angle of reflection of the reflected light becomes $2\Delta\theta$, and the corresponding linear displacement Δx of the reflected light on the detector surface at a distance L can be expressed as Eq. (3-9). $\Delta\theta$ can thus be obtained by placing an image sensor on the detector surface and detecting Δx. This method is also referred to as the optical lever method. It should be noted that L should be kept constant while the angular displacement measurement since

わるため，L を一定に保つ必要がある．

　この方法に対して焦点距離 f を有するレンズを導入し，距離 L によらない $\Delta\theta$ の検出を実現した角度センサがオートコリメータである（Fig. 3-6）．光源からの測定光に照射されたターゲットの像は，導入したレンズによってコリメート光となって測定対象に搭載した反射鏡で反射されたのち，再度このレンズを通過して受光素子上で像を結ぶ．このとき，測定対象に角度変位 $\Delta\theta$ が発生すると，式(3-10)で示される Δd だけ受光素子上の像がシフトする．なお，Δd は測定対象と受光素子との間の距離 L に影響を受けない．この Δd を CCD，CMOS などのイメージセンサや 1 次元/2 次元 PSD などの位置センサで検出することで，$\Delta\theta$ を高精度に検出できる．光源にはキセノンランプや LED などの白色光源，または指向性に優れた半導体レーザ光源や超短パルスレーザ光源（第 14 章）が用いられる．

小型レーザオートコリメータ　レーザ光源を採用し，受光素子として分割型フォトダイオード（PD）を用いたレーザオートコリメーション法に基づく角度センサでは，イ

Δx could be affected by L.

　By introducing a lens with the focal length f into this method, angular displacement measurement independent of distance L can be achieved. Fig. 3-6 shows a schematic of the optical setup referred to as the autocollimator. The image of the target irradiated by the measurement light from the light source is collimated by the introduced lens, reflected by a reflector mounted on the measurement target, and then passes through the lens again to form an image on the photodetector. When an angular displacement $\Delta\theta$ is generated in the measurement target, the image on the photodetector will be shifted by Δd which can be calculated as Eq. (3-10). It should be noted that Δd will not be affected by L, as can be seen in the equation. $\Delta\theta$ can thus be detected with high accuracy by detecting Δd with an image sensor such as a charge-coupled device (CCD) or a complementary metal-oxide-semiconductor (CMOS) sensor or a position-sensitive detector such as a one or two-dimensional PSD. White light sources such as xenon lamps and LEDs, or semiconductor laser sources and ultrashort pulse laser sources (Chapter 14) with good directivity are employed as light sources.

A compact laser autocollimator　Laser autocollimation method with a laser source and a segmented photodiode (PD) can realize highly sensitive angular displacement measurement in a compact form, although the angular measurement

メージセンサを用いたオートコリメータに比べて角度測定レンジは制約を受けるもの
の，感度の高い角度変位検出をコンパクトな形態で実現できる．Fig. 3-7 に，本手法
に立脚したレーザオートコリメータ光学系の模式図を示す．測定面にはビーム径一定
のコリメートレーザ光を照射し，その反射光をレンズで PD 上に集光する．分割型 PD
上の各素子からの光電流の割合をもとに，集光スポットの変位を高感度に得ることが
できる．4 分割 PD を用いた場合，2 自由度の角度 $\Delta\theta_X$, $\Delta\theta_Y$ を同時に検出可能である．
この光学系では，受光素子上における集束レーザ光の強度分布が一定であるとみなせ
る場合，角度検出感度はレンズの焦点距離にほぼ無関係となることから，短焦点レン
ズを適用して読み取りヘッドをコンパクトに構成しつつ，高感度な角度変位検出が可
能である．

3 軸レーザオートコリメータ　レーザオートコリメーション法に基づくレーザオート

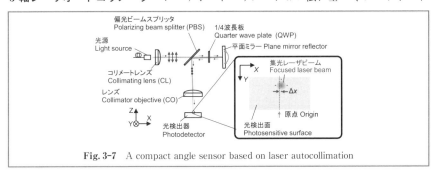

Fig. 3-7　A compact angle sensor based on laser autocollimation

range will be restricted than that of an autocollimator using an image sensor.
Fig. 3-7 shows a schematic diagram of the laser autocollimator based on this method. A collimated laser beam with a constant beam diameter is irradiated onto the
measurement surface, and the reflected light is focused onto the PD by a lens. The
displacement of the focused spot can be obtained with high sensitivity based on the
ratio of the photocurrent from each element on the segmented PD. Two de-
gree-of-freedom angular displacements $\Delta\theta_X$ and $\Delta\theta_Y$ can be detected simultaneously
when a four-element PD is used. In this optical system, the angle detection sensitivity is almost independent of the focal length of the lens under the case where the
intensity distribution of the focused laser beam on the photosensor could be considered to be constant. High-sensitivity angular displacement measurement could thus
be possible with a compact reading head having a lens with a short focal length.

A three-axis laser autocollimator　A three-axis laser autocollimator can be realized by employing a diffraction grating as a measurement target on behalf of a flat

コリメータにおいて，ターゲットとして回折格子を用い，0 次反射光とともに ±1 次
反射回折光を利用することで，従来の 2 軸レーザオートコリメータでは原理的に検出
できなかったレーザ光軸周りの微小回転 $\Delta\theta_Z$ も検出可能となる．Fig. 3-8 に光学系の
模式図を示す．測定レーザ光が，その波長 λ に対して大きいピッチ g を有する回折格
子に対して垂直に入射した場合，0 次反射光とともに ±1 次反射回折光が発生する．0
次反射光は $\Delta\theta_Z$ によって影響を受けないが，±1 次回折光は受光素子上での集光位置
が変化する．この集光位置の変動から，演算式をもとに $\Delta\theta_Z$ も検出可能となる．なお，
$\Delta\theta_X$ および $\Delta\theta_Y$ は 0 次反射光をもとに従来手法どおりに検出可能である．

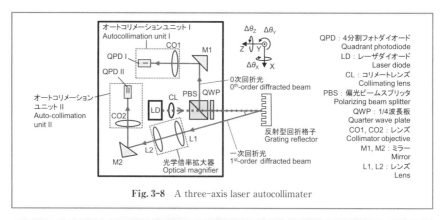

Fig. 3-8　A three-axis laser autocollimater

mirror reflector in conventional laser autocollimator. In the three-axis laser autocollimator, the positive and negative first-order diffracted beams from the target grating are used together with the zeroth-order diffracted beam (reflected light). This
enables the detection of small angular displacement $\Delta\theta_Z$ about the laser axis, which
cannot be detected by conventional two-axis laser autocollimators in principle.
Fig. 3-8 shows a schematic diagram of the optical system. When the measurement
laser beam is projected onto the grating target with a large pitch g relative to the
wavelength λ at a right angle, the first-order diffracted beams are generated, along
with the zeroth-order reflected beams which will not be affected by $\Delta\theta_Z$. The variations of the focusing positions of the first-order diffracted beams induced by $\Delta\theta_Z$
make it possible to detect $\Delta\theta_Z$ based on the arithmetic operation. It should be noted
that $\Delta\theta_X$ and $\Delta\theta_Y$ can be detected in the same manner as the conventional two-axis
laser autocollimator by using the the zeroth-order reflected beam that will not be
affected by $\Delta\theta_Z$ but will be affected by $\Delta\theta_X$ and $\Delta\theta_Y$.

【演習問題】

3-1) 長さ 100 mm のサインバーで角度 28° の基準を生成するのに必要なブロックゲージの組み合わせを求めよ．

3-2) アブソリュート型ロータリエンコーダにおいて，角度分解能 0.02 角度秒を実現するのに必要となる，バイナリコードのビット数を求めよ．

3-3) 焦点距離 100 mm のレンズを用いたレーザオートコリメータにおいて，分解能 0.01 角度秒を実現するのに必要な受光素子の変位検出分解能を求めよ．

【Problems】

3-1) Find the combination of gauge blocks required to generate a reference angle of 28° with a sine bar of length 100 mm.

3-2) Find the number of bits of binary code required to achieve an angular resolution of 0.02 arc-second for an absolute rotary encoder.

3-3) Find the displacement detection resolution of a photodetector required to achieve a resolution of 0.01 arc-second in a laser autocollimator using a lens with a focal length of 100 mm.

第4章　時間スケール

　物理的な現象のうち時間的に変化する過渡現象には様々な場面で遭遇する．この過渡現象は，物理現象のプロセスを理解するためだけでなく，フィードバック制御などの機器制御にも利用できる．この過渡現象を精密に測定するためには正確な時間軸のスケールが必要であり，第1章でも述べたとおり，時間を周波数として捉え直すことで実現している．言い換えると，精密に繰り返される既知の周期的な現象を正確に計数することで時間スケールとして用いている．これまでに，この周期現象には，水晶振動子や音叉，発振回路などを利用してきた．また，外乱の影響を受けにくい原子や分子の振動現象を周波数発生技術としても利用している．例えば時間の定義は，従来は「地球の自転（1 太陽日の 86 400 分の1）」や「公転（1 太陽年の 31 556 925.974 7 分の1）」のような天体活動を基準とした時間を利用していたが，1967 年の第 11 回国際度量衡委員会において，「^{133}Cs の共鳴マイクロ波の周期の 9 192 631 770 倍」と原子固

Chapter 4　Time Scale

　Time-varying transient phenomena are encountered in a variety of physical phenomena. These transients can be used not only to understand the process of a physical phenomenon, but also for feedback control of equipment. Precise measurement of these transients requires an accurate time scale, which is achieved by redefining time as frequency. In other words, it consists of accurately counting the number of occurrences of a known periodic phenomenon that is precisely repeated on a time scale. Until now, quartz crystals, tuning forks, and oscillation circuits have been used for this periodic phenomenon. Frequency generation techniques have also utilized vibrational phenomena of atoms and molecules, which are less susceptible to external disturbances. For example, the definition of time has traditionally been based on celestial activities such as the earth's rotation (1/86 400 of a solar day) and orbit (1/31 556 925.974 7 of a solar year), but in 1967, the 11th International Committee of Weights and Measures adopted the definition of time as "the period of

有の周期現象を基準として採用していることからも，周期現象が，正確な時間スケールを実現する上で必要不可欠であることがわかる．本章では，正確な時間スケールを確立するための周波数測定技術として，周波数発生技術と周波数測定技術について述べる．

● 4.1 水晶時計・原子時計

正確な時間スケールの実現には，共鳴現象などの精密に繰り返される既知の周期現象を標準周波数発生器として参照利用する．一般的には，安定性の高い水晶製の振動子を共振回路に導入した水晶時計や，さらに精密にするための原子の共鳴周波数を利用した原子時計がある．本節では，水晶振動子の振動原理および原子の共鳴現象について述べる．

水晶時計は，結晶に電圧を印加した際に生じる変形現象である逆圧電効果を利用した振動子である．水晶は，化学的および物理的に安定な SiO_2 の六方晶系の結晶である．圧電的および弾性的に異方性を示すため，Fig. 4-1 に示すとおり，結晶の切り出し方により特性が異なる．一般的には，温度安定性に優れる AT カットや GT カット

a resonant microwave of [133]Cs 9 192 631 770th of the period of a resonant microwave of [133]Cs" and the fact that atomic clocks were used as the standard indicates that periodic phenomena are indispensable to the time scale. This chapter describes frequency generation and frequency measurement techniques accurately determining the time scale.

● 4.1 Crystal and atomic clocks

Accurate time scales are achieved by using known periodic phenomena that repeat precisely, such as resonance phenomena, as reference standard frequency generators. Generally, there are quartz clocks that use highly stable quartz crystals in resonant circuits, and atomic clocks that use resonant frequencies of atoms to achieve even greater precision. This section describes the principle of quartz crystal oscillation and the phenomenon of atomic resonance.

Quartz crystal units utilize the reverse piezoelectric effect, a deformation phenomenon that arises due to the application of a voltage to the crystal. Quartz is a hexagonal crystal of SiO_2 that is chemically and physically stable. Since it is piezoelectrically and elastically anisotropic, as shown in Fig. 4-1, its properties vary de-

と呼ばれている方向に切り出された結晶が水晶時計に用いられている．これらの結晶
に通電させるための金蒸着を行い，交流電圧を印加することにより，それぞれ，滑り
振動および縦振動が発生する．これらの振動には，結晶の形状により固有振動数が存
在し，印加する交流電圧の周波数が固有振動数になると共振現象により電流が増大す
る．このとき，水晶振動子は，リアクタンスとして作用するため，Fig. 4-2に示すよう
に，等価回路で置き換えることができることが知られている．水晶振動子のインピーダ
ンスの共振周波数は，直列共振に相当するf_sおよび並列共振に相当するf_pが存在する
が，コイルとして作用する$f_s < f < f_p$の範囲で使用する．MHz以下で利用されることが
多く，不確かさは10^{-5}程度である．それ以上の高精度化には原子時計を用いる．

^{133}Csを利用した原子時計の原理をFig. 4-3に示す．セシウム原子には，超微細構造

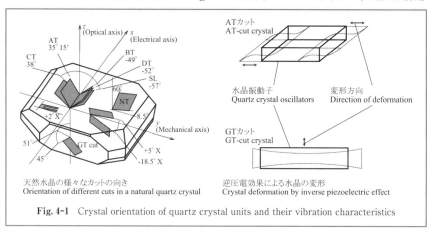

天然水晶の様々なカットの向き
Orientation of different cuts in a natural quartz crystal

逆圧電効果による水晶の変形
Crystal deformation by inverse piezoelectric effect

Fig. 4-1　Crystal orientation of quartz crystal units and their vibration characteristics

pending on the way the crystal is cut. Generally, AT-cut and GT-cut crystals are
used because of their excellent temperature stability. Gold deposition to conduct
current through these crystals and the application of an alternating voltage generate sliding vibrations and longitudinal vibrations, respectively. Depending on the
shape of the crystal, there are natural frequencies, and at a certain frequency, the
current increases due to resonance phenomena. Since the crystal unit acts as a re-
actance, it can be replaced by an equivalent circuit as shown in Fig. 4-2. The resonant frequency of the impedance of the crystal unit has f_s corresponding to series
resonance and f_p corresponding to parallel resonance, but it is used in the range of
$f_s < f < f_p$, where it acts as a coil. For higher accuracy, atomic clocks are used.

The principle of the atomic clock using ^{133}Cs is shown in Fig. 4-3. The cesium
atom has many energy levels called hyperfine structures, and in the ground state,

と呼ばれるエネルギー準位が多数存在し，基底状態では原子の持つ全角運動量に対して２つの基底状態が存在する．さらに，各基底状態には，磁気スピンに依存した複数のエネルギー状態が存在し，各基底準位の磁場安定性に優れた準位同士の遷移周波数が標準周波数となる．約 100℃で熱せられたセシウムは，両基底状態が混在した状態であるが，磁石により一方の基底状態のみがラムゼー共振器に導入される．ラムゼー共振器では，ある時間間隔 T で２回に分けてマイクロ波を照射することにより，量子遷移の発生確率が周波数に敏感になる．このとき，遷移確率分布は，式(4-3)で示されるようにマイクロ波の周波数 ω と遷移周波数 ω_0 の差分が 0 のときに最大の確率とな

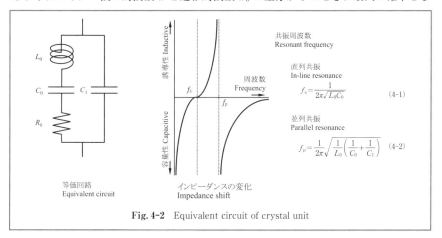

Fig. 4-2 Equivalent circuit of crystal unit

there are two ground states for the total angular momentum of the atom. Each ground state has several energy states depending on the magnetic spin, and the standard frequency is the transition frequency between levels with excellent magnetic field stability. Cesium heated at about 100° has a mixture of both ground states, but only one of the ground states is introduced into the Ramsey resonator by the magnet. In the Ramsey resonator, the probability of quantum transitions becomes frequency-sensitive by irradiating microwaves twice with a certain time interval T. The transition probability distribution is maximized when the difference between the microwave frequency ω and the transition frequency ω_0 is zero, as shown in Eq. (4-3), and the standard frequency is obtained by feedback control. Since the half-width is $1/2T$, the wider the distance between the two microwave irradiation sources, the more accurately the measurement can be performed. Note that μ_B is the Bohr magneton and B is the magnetic flux density. τ is the microwave

るためフィードバック制御することで標準周波数を得る．また，その半値幅は $1/2T$ となるため，2 つのマイクロ波照射源の間隔が広い方が高精度に測定することができる．なお，μ_B はボーア磁子，B は磁束密度である．また，τ はマイクロ波の照射時間，l は照射距離，L はラムゼー共振器の共振器長であり，$T/\tau=L/l$ の関係が成り立つ．原子遷移を選別する方法として，他にもレーザ冷却を利用した原子泉方式など様々な方法がある．さらに，原子時計には，セシウム以外にも，アンモニア原子，水素メーザやルビジウム原子を用いたものがある．また，光格子時計のようなさらに高精度な方法も提案されている．

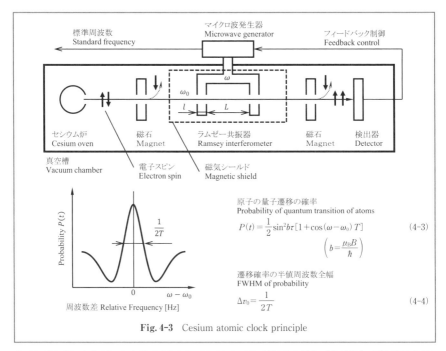

原子の量子遷移の確率
Probability of quantum transition of atoms

$$P(t)=\frac{1}{2}\sin^2 b\tau[1+\cos(\omega-\omega_0)\,T]\tag{4-3}$$

$$\left(b=\frac{\mu_B B}{\hbar}\right)$$

遷移確率の半値周波数全幅
FWHM of probability

$$\Delta v_0=\frac{1}{2T}\tag{4-4}$$

Fig. 4-3　Cesium atomic clock principle

irradiation time, l is the irradiation distance, L is the resonator length of the Ramsey resonator and the relation $T/\tau=L/l$ holds. There are various other methods for selecting atomic transitions, such as the atomic spring method using laser cooling, but the method currently used as the national primary frequency standard achieves an uncertainty of 10^{-16}. In addition to Celsium, other types of atomic clocks use ammonia, hydrogen maser, and rubidium atoms. Even higher precision methods such as optical lattice clocks have also been proposed.

● 4.2　周波数発生器

機器の制御や計測に正弦的に振動する信号を利用することはよくあることであるが，その際に，振動周期の正確さが制御もしくは測定精度に大きく影響することがある．そのため，正確に周波数制御された正弦波信号を得ることは重要である．一般的に，発生させた信号の周波数を電気回路的に補正することで所望の周波数信号を得る．この補正には，基準となる信号と正弦波信号の位相を比較することで実現している．具体的には，Fig. 4-4 に示す位相同期検波方法で行う．まず，基準となる信号 f_{in} を分周器で任意の周波数に分周したのちに位相同期回路（PLL）に入力する．PLL では，まず入力信号と出力信号の位相を位相比較器で比較し，位相のずれに相当する電流が出力される．この電流は，ループフィルタで DC 電圧として出力される．その後，電圧制御発振器（VCO）で補正された周波数を有する信号が発生し，分周器で分周された後にフィードバック制御を行うことで安定した周波数信号を得ることができる．な

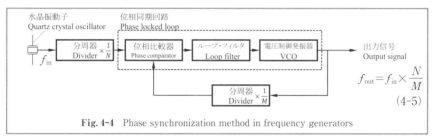

Fig. 4-4　Phase synchronization method in frequency generators

● 4.2　Frequency generator

Sinusoidal oscillating signals are often used to control or measure equipment, and the accuracy of the oscillation period can greatly affect the accuracy of control or measurement. Therefore, it is important to obtain a precisely frequency-controlled sinusoidal signal. Generally, the desired frequency signal is obtained by electrically correcting the generated signal. This correction is achieved by comparing the phase of the sinusoidal signal with a reference signal. Specifically, the phase-locked detection method shown in Fig. 4-4 is used. First, the reference signal f_{in} is divided by a frequency divider to a desired frequency and then input to a phase-locked loop (PLL). The phase comparator in the PLL which compares the phase of the input and output signals and outputs a current corresponding to the phase shift. This current is output as a DC voltage by a loop filter. The voltage-controlled oscillator (VCO) then generates a signal with a corrected frequency, which is divided by the frequency divider and then feedback-controlled to obtain a stable frequency signal. Since the input and output signals must be in phase, the loop gain must be greater

お，入力信号と出力信号が同位相である必要があるため，ループ・ゲインが1以上であることが発振の条件である．

PLL を構成する VCO は，入力電圧を制御信号として任意の周波数を得ることができる発振器であり様々な方式が提案されている．その一例として，Fig. 4-5 に，コルピッツ型 LC 発振回路に可変コンデンサを導入した VCO を示す．LC 型発振回路では，インダクタンス素子（L）と充電状態にあるキャパシタンス素子（C）を並列につなぐことで，C の電荷が L を経て一定の周波数で振動しながら減衰するが，減衰分を外部より補うことで発振回路を構築できる．コルビッツ型 LC 発振回路は，コレクタとベース間に2つのコンデンサを直列に接続し，2つのコンデンサの接点をエミッタに接続することで，出力同調信号の一部を入力側へ帰還させる方式の発振回路である．発振周波数は，C_1 および C_2 を変更することで制御することができる．しかしながら，制

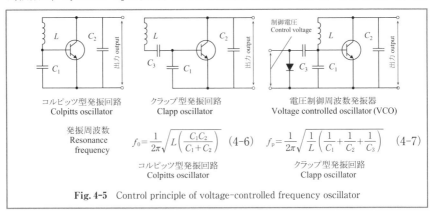

Fig. 4-5　Control principle of voltage-controlled frequency oscillator

than or equal to 1 for oscillation to occur.

The VCO that constitutes the PLL is an oscillator that can obtain an arbitrary frequency by using the input voltage as a control signal, and various methods have been proposed. As an example, Fig. 4-5 shows a VCO in which a variable capacitor is introduced into a Colpitts-type LC oscillation circuit. An oscillation circuit can be constructed by supplementing the attenuation externally. The Colpitts-type LC oscillation circuit is an oscillation circuit in which two capacitors are connected in series between the collector and base, and the contacts of the two capacitors are connected to the emitter to return part of the output tuned signal to the input side. The oscillation frequency can be controlled by changing C_1 and C_2. However, because the control is complicated, the Clapp-type oscillation circuit, which is a development of the Korbitz-type circuit, is often used. In the Clapp-type oscillation cir-

御が煩雑となるため，コルピッツ型を発展させたクラップ型発振回路を利用すること
が多い．クラップ型発振回路では，C_3 に可変コンデンサを用いることで任意の周波数
を設定することができる．このクラップ型発振回路の周波数を電圧制御できるように
したのが VCO である．VCO では，C_3 の一端に可変コンデンサと入力端子を設けるこ
とで電圧により周波数が可変できる構成としている．この入力信号として，ループ・
フィルタからの位相差信号を入力することで設定周波数に正確に同調した周波数を得
ることができる．

● 4.3　周波数カウンタ

　信号の周波数は，同じ周波数を有するパルス波に変換し，このパルス数を基準周波
数より分周もしくは逓倍して得た周波数で周期的に所定の時間ゲート内のパルス数を
計数することで実現している．Fig. 4-6 に，周波数カウンタの基本構成を示す．入力
信号は，ある程度の強度まで増幅回路により増幅された後に，波形整形回路で方形波
に整形される．その後，微分回路で時間的なエッジが抽出され，ゲート時間内のパル
ス数を計数器でカウントする．したがって，基準発振器の性能は，装置の精度に直接
影響を与えるため，正確な発振器が必要である．一般的には，水晶振動子が用いられ

cuit, C_3 can be set to any frequency by using a variable capacitor. A VCO allows
the frequency of this clapped oscillation circuit to be controlled by a voltage, and is
composed of a variable capacitor and an input terminal at one end of C_3. A phase
difference signal from a loop filter can be input as the input signal to obtain a frequency that is precisely in tune with the set frequency.

● 4.3　Frequency counter

The frequency of the signal is realized by converting it into a pulse wave having
the same frequency and counting the number of pulses within a predetermined
time gate periodically at a frequency obtained by dividing or multiplying this pulse
number by the reference frequency. Fig. 4-6 shows the basic configuration of the
frequency counter. The input signal is amplified to a certain level of strength by an
amplifier circuit and then shaped into a square wave by a waveform shaping circuit. The temporal edges are then extracted by a differentiation circuit, and the
number of pulses within the gate time is counted by a counter. Therefore, the performance of the reference oscillator directly affects the accuracy of the device, re-

るが，さらに安定度が必要となればルビジウム原子時計などを用いることもある．

● 4.4　RF スペクトラムアナライザ

　信号に含まれる周波数やその強度を測定する機能を有するのが RF スペクトラムア
ナライザである．その測定方法には，バンドパスフィルタや FFT を利用した方法が
あるが，高精度に測定できる方法にヘテロダイン式の掃引同調型の測定方法がある．
Fig.4-7 に，ヘテロダイン式スペクトラムアナライザの測定原理を示す．入力信号は，
EF 減衰器とローパスフィルタにより，ノイズや高調波などの不要な成分が除外され
る．一方で，基準周波数で校正された周波数発生器からの所定の周波数を有する信号

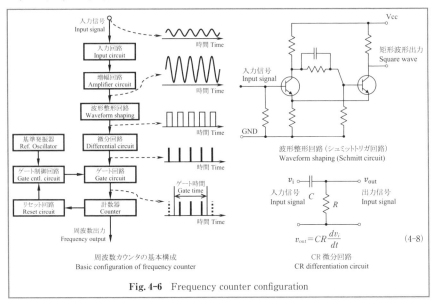

Fig. 4-6　Frequency counter configuration

quiring an accurate oscillator. Generally, quartz crystals are used, but rubidium
atomic clocks are sometimes used when greater stability is required.

● 4.4　RF spectrum analyzer

　Spectrum analyzers have the function of measuring the frequencies contained in
signals and their intensities. Fig. 4-7 shows the measurement principle of a heterodyne spectrum analyzer. The input signal is attenuated by an RF attenuator and a
low-pass filter to eliminate unwanted components such as noise and harmonics. On
the other hand, the signal is multiplied by a mixer with a signal with a specified
frequency from a frequency generator calibrated at the reference frequency. As

とミキサーで乗算される．このとき，式(4-9)で示したとおり，差周波と和周波で混合
された波が得られる．一般的には，差周波を利用することが多く，差周波増幅器で振
幅が増幅された後にロブアンプで包絡線を得る．最終的にこの包絡線から所定の周波
数の振幅を得る．周波数発生器の周波数は掃引発生器により連続的に掃引されるので
上記の検出を繰り返すことで各周波数ごとの振幅強度を得ることができる．

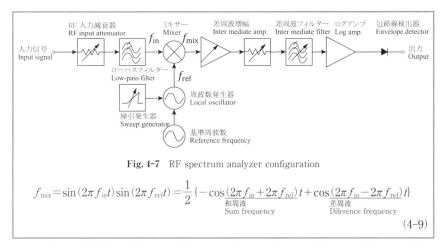

Fig. 4-7　RF spectrum analyzer configuration

$$f_{\mathrm{mix}} = \sin(2\pi f_{\mathrm{in}} t)\sin(2\pi f_{\mathrm{ref}} t) = \frac{1}{2}\{-\cos\underbrace{(2\pi f_{\mathrm{in}} + 2\pi f_{\mathrm{ref}})}_{\substack{\text{和周波}\\ \text{Sum frequency}}} t + \cos\underbrace{(2\pi f_{\mathrm{in}} - 2\pi f_{\mathrm{ref}})}_{\substack{\text{差周波}\\ \text{Diference frequency}}} t\}$$

(4-9)

shown in Eq. (4-9), a wave mixed with the difference frequency and the sum frequency is obtained. Generally, the difference frequency is used in most cases, and
the amplitude is amplified by a difference frequency amplifier, and then the envelope is obtained by a lob amplifier. Finally, the amplitude of a given frequency is
obtained from this envelope. Since the frequency of the frequency generator is
swept continuously by the sweep generator, the amplitude intensity for each frequency can be obtained by repeating the above detection.

【演習問題】

4-1) 水晶振動子の AT カットと GT カットの共振周波数をそれぞれ調べよ.

4-2) セシウム原子時計の標準周波数を調べよ.

4-3) 式(4-8)の微分方程式を解くことにより v_{out} の時間的な挙動を考察せよ.

4-4) バンドパスフィルタや FFT を利用したスペクトラムアナライザの動作原理について調べよ.

【Problems】

4-1) Find the resonant frequencies of quartz crystals by AT cut and GT cut, respectively.

4-2) Show the standard frequency of a cesium atomic clock.

4-3) Solve the differential equation of Eq. (4-8) to investigate the time behavior v_{out}.

4-4) Investigate the principle of operation of a spectrum analyzer using a bandpass filter or FFT.

第5章　幾何形状と表面性状の計測

寸法，サイズ，幾何形状，表面性状の計測と評価は，ものづくりや製造現場における生産管理の計測項目の中でかなりの比率を占めるため，非常に重要である．本章では，幾何形状と表面性状の計測について焦点を当てる．

● 5.1　幾何公差と表面性状・表面粗さ

製造現場での測定と検査には様々な項目が存在するが，幾何学量において重要な項目には，寸法の公差（旧・寸法公差），サイズ公差，幾何公差，表面性状が挙げられる．これらの項目を測定する機器としては，ノギス，マイクロメータ，測長機，測定顕微鏡，三次元座標測定機，光学式三次元スキャナ，真円度測定機，輪郭形状測定機，触針式表面粗さ測定機，光学式表面性状測定機などがある．本節では，前述の公差の

Chapter 5　Measurement of Geometrical Form and Surface Texture

Measurement and evaluation of dimension, size, geometrical form and surface texture are crucial because they represent a substantial proportion of the physical properties that require measurements in production control at manufacturing and production sites. This chapter focuses on the measurement of the geometrical form and the surface texture.

● 5.1　Geometrical tolerances and surface texture/roughness

Among the various measurement and verification specifications at the manufacturing site, the most important ones in macrogeometry and microgeometry are dimensional tolerance, size tolerance, geometrical tolerance, and surface texture (surface roughness, waviness, and lay). Measurement instruments for measuring these specifications include vernier caliper, micrometer, dial indicator, length measuring machine, measuring microscope, coordinate measuring machine (CMM), 3D scanner, industrial X-ray CT inspection system, roundness measurement instrument, contour measurement instrument, surface roughness measurement instrument, optical surface texture measurement instrument, etc. This section describes geometri-

中でも幾何公差，および，表面性状・表面粗さについて解説する．

　Table 5-1 は，JIS B 0021（ISO 1101）の幾何公差の種類とその特性，および，記号である．まず，幾何公差については，Table 5-1 のうち平面度について説明する．平面度とは，理論的に正確な平面形体（TEF）からの実形体の狂いの大きさのことである．Fig. 5-1 は，平面度を図示した例である．Fig. 5-1 の仕様で定義される公差域は，Fig. 5-2 のように距離 *t* だけ離れた平行 2 平面によって規制される．公差域とは，公差（1 つ以上の長さ寸法による特性），および，1 つまたは 2 つの理想的な線または面によって規制される領域のことである．以上より，Fig. 5-1 の図示記号は，製品の表面が 0.06 mm（60 μm）だけ離れた平行 2 平面の中に存在しなければならないことを意味する．この意味について平易な表現をすれば，「幾何公差で指示された位置の形状偏

Table. 5-1　Symbols for geometrical characteristics

Specification 公差の種類	Characteristics 特性	Symbol 記号	Datum needed データム指示	Specification 公差の種類	Characteristics 特性	Symbol 記号	Datum needed データム指示
Form 形状公差	Straightness 真直度	—	no 否		Position 位置度	⊕	yes・no 要・否
	Flatness 平面度	▱	no 否	Location 位置公差	Concentricity (for centre points) 同心度 (中心点に対して)	◎	yes 要
	Roundness 真円度	○	no 否				
	Cylindricity 円筒度	⌀	no 否		Coaxiality (for median lines) 同軸度 (軸線に対して)	◎	yes 要
	Line profile 線の輪郭度	⌒	no 否				
	Surface profile 面の輪郭	⌓	no 否		Symmetry 対称度	⩵	yes 要
Orientation 姿勢公差	Parallelism 平行度	//	yes 要		Line profile 線の輪郭度	⌒	yes 要
	Perpendicularity 直角度	⊥	yes 要		Surface profile 面の輪郭	⌓	yes 要
	Angularity 傾斜度	∠	yes 要	Run-out 振れ公差	Circular run-out 円周振れ	↗	yes 要
	Line profile 線の輪郭度	⌒	yes 要		Total run-out 全振れ	⌀↗	yes 要
	Surface profile 面の輪郭	⌓	yes 要				

cal tolerance and surface texture/roughness as two of these important terms.

　Table 5-1 presents the specifications of the geometrical tolerances of ISO 1101, with their characteristics and symbols. Among the geometrical characteristics listed in Table 5-1, flatness is explained below. Flatness is the form error deviation of the real feature from the planar feature, which is the theoretically exact feature (TEF). Fig. 5-1 shows an example of flatness indication. The tolerance zone defined by the specification in Fig. 5-1 is limited by two parallel planes a distance *t* apart as shown in Fig. 5-2. The tolerance zone is a space limited by and including one or two ideal lines or surfaces, and characterized by one or more linear dimensions, called a tolerance. From the above, the indications and symbols in Fig. 5-1 mean that the surface of the product shall be contained between two parallel planes 0.06 mm（60 μm）apart. In other words, the meaning of these indications and symbols is "the form deviation, irregularities, and/or smoothness on the location indicated by the geomet-

差や凸凹，平坦さは，0.06 mm（60 μm）以下でなければならない」となる．

次に，微細形状である表面性状について解説する．表面性状とは，物体の表面にある微細な幾何学的凹凸のことである．表面性状の測定・評価を最も平易な言葉で言い表せば，ざらざら，つるつる，でこぼこ，ぴかぴかの程度を定量的に測定・評価することである．JIS/ISO の規格群は，2 次元：線（輪郭曲線方式）と 3 次元：面領域の表面性状に大別されるが，本章では輪郭曲線方式に焦点を当てる．

輪郭曲線方式の JIS/ISO 規格における表面性状の評価は，大きく分けて 3 つの曲線：断面曲線，粗さ曲線，うねり曲線によって行われる．これらは，波長領域によって示すと Fig. 5-3 のようになり，形状誤差，うねり，粗さ，および，ノイズや量子化誤差などの微細な成分に分離処理される．断面曲線（P 曲線）とは，表面を測定する

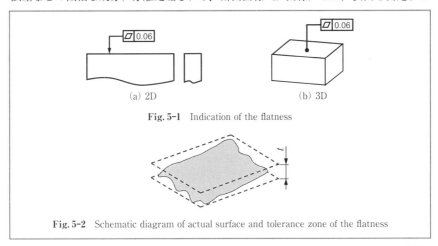

(a) 2D   (b) 3D

Fig. 5-1   Indication of the flatness

Fig. 5-2   Schematic diagram of actual surface and tolerance zone of the flatness

rical tolerance, shall be 0.06 mm（60 μm）or less".

Next, the surface texture, which is microgeometry, is explained. The surface texture is the term used to describe geometrical irregularities on the surface of a workpiece. The measurement and evaluation of surface texture can be expressed in its simplest terms as the quantitative measurement and evaluation of the degree of asperous, slickness, smoothness, luster, or bumps and dips. The JIS/ISO standards are roughly divided into 2D, which is the surface texture traced by a single line (profile method), and 3D, which is the areal surface texture, then this chapter focuses on the profile method.

In the JIS/ISO standards for the profile method, the surface texture evaluation is performed using three broadly divided profiles: primary profile, roughness profile, and waviness profile. These, when considered in the wavelength domain, are represented as shown in Fig. 5-3, and are filtered by numerical calculation into form error,

ことで得られた機械的曲線からの形状誤差の除去と，カットオフ λ_s のローパス特性の輪郭曲線フィルタ（S フィルタ）による粗さ成分より短い波長成分の除去とを実施した曲線である．粗さ曲線（R 曲線）は，断面曲線から粗さ成分よりも長い波長成分をカットオフ λ_c のハイパス特性の輪郭曲線フィルタ（L フィルタ）によって除去した曲線である．うねり曲線（W 曲線）は，カットオフ λ_c のローパス特性の輪郭曲線フィルタ（S フィルタ），および，カットオフ λ_f のハイパス特性の輪郭曲線フィルタ（L フィルタ）によって帯域通過させた曲線である．

　以上のフィルタ処理によって得られたそれぞれ 3 つの曲線から，表面性状パラメータである断面曲線パラメータ（P パラメータ），うねりパラメータ（W パラメータ），および，粗さパラメータ（R パラメータ）が計算される．各曲線の表面性状パラメータの記号は，どの曲線のパラメータであるかを明確にするために，断面曲線の場合は

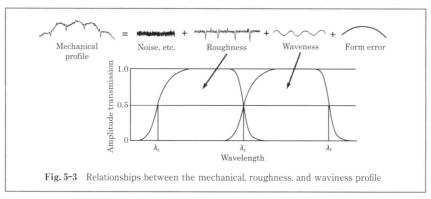

Fig. 5-3　Relationships between the mechanical, roughness, and waviness profile

waviness, roughness, and very small components such as noise and quantization error. In ISO 21920-2, the primary profile (P-profile) is a scale-limited profile derived from the mechanical profile by removing small-scale lateral components using a profile S-filter with a low-pass characteristic of a cut-off λ_s and then removing the form error using a profile F-operation. The roughness profile (R-profile) is a scale-limited profile derived from the primary profile by removing large-scale lateral components by a specific type of profile L-filter with a high-pass characteristic of cut-off λ_c. The waviness profile (W-profile) is a scale-limited profile derived from the primary profile by removing both small-scale lateral components by the profile S-filter of cut-off λ_c and large-scale lateral components by the profile L-filter of cut-off λ_f.

　From the three profiles obtained by the above filtering process, the surface texture parameters: P-parameter (from the primary profile), W-parameter (from the waviness profile) and R-parameter (from the roughness profile) are calculated. The first capital letter for the symbol of the surface texture parameter indicates "P"

P, 粗さ曲線の場合は R, うねり曲線の場合は W から始まるパラメータとなっており, その次の記号がパラメータの幾何学的, 統計的な意味を表している. 例えば, 算術平均高さ Pa, Wa, Ra の a は arithmetical を表しており, 計算は輪郭曲線の高さ方向の絶対値平均を行う. また, 平均長さ Psm, Wsm, Rsm の sm は輪郭曲線要素の幅の平均値を表しており, 表面の凹凸の平均波長を特徴付けることができる.

　最後に, 図示記号によって製品技術文書 (例えば, 図面, 仕様書, 契約書, 報告書, および, 関連文書など) に表面性状の要求事項を指示する方法を説明する. Fig. 5-4 は, 輪郭曲線において基本となる図示記号である. Fig. 5-4 に示すように, 図示記号は加工方法の要求によって 3 種類あり, 記号の上のバーは 2002 年版の ISO 1302 (2003 年版の JIS B 0031) で規定されていた図示記号と, 2016 年版の ISO 25178-1 (2023 年

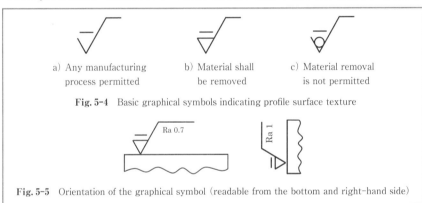

a) Any manufacturing　　b) Material shall　　c) Material removal
   process permitted　　　　be removed　　　　is not permitted

Fig. 5-4 Basic graphical symbols indicating profile surface texture

Fig. 5-5 Orientation of the graphical symbol (readable from the bottom and right-hand side)

for Primary profile, "R" for Roughness profile, and "W" for Waviness profile in order to clarify which profile the parameter is for, and the following letter (s) indicate the geometric and statistical meaning of the parameter. For example, "a" in the arithmetic mean height Pa, Wa, and Ra means for "arithmetic," and the calculation for Pa, Wa, and Ra is performed by averaging the absolute values in the height direction of each profile. In addition, "sm" of the mean profile element spacing Psm, Wsm, and Rsm means the mean value of the profile element spacings and can characterize the mean value of the width of the surface irregularities.

　Finally, graphical symbols and rules for the indication of profile surface texture specifications in technical product documentation (TPD) are briefly described, where TPD is technical drawing and also includes, for example, technical product specification, contract document, report, and related documentation, etc. Fig. 5-4 shows the basic graphical symbol or indication of surface texture by profile. As shown in Fig. 5-4, the basic graphical symbol is defined in three types depending on the requirements for the production process, and the bar above the symbols is used

版の JIS B 0681-1）の三次元の表面性状の図示記号とを区別するためのものである．図示記号の向きと位置に関しては，ISO 129-1（JIS Z 8317-1）の規定に従い Fig. 5-5 に示すように図面の下側または右側から図示記号が読めるように指示する．また，表面性状の図示記号は，円筒形体を例外として，寸法線から離れた位置にだけ指示できる．

● 5.2　移動平均法

幾何形状や表面性状の測定データは，系列データであることが多い．系列データの代表的なものとしては時系列データが挙げられ，これはある現象の時間的な変化を測定して得られた連続の値のことである．物体の形状に関しては，位置に依存した変化を連続して測定することで，幾何学的な変化の系列データが得られる．系列データから有益な情報を得るためには，統計的手法やデータサイエンスの手法などを適用する必要がある．これらの手法の中で基礎的かつ重要な手法としては移動平均法が挙げられるため，本節ではこの移動平均法について解説する．

to distinguish these symbols from the ones specified in ISO 1302 and those specified for indication of areal surface texture specifications in ISO 25178-1. For the orientation and position of the graphical symbol on the drawing, the graphical symbol shall be indicated so that it can be read from the bottom or right-hand side of the drawing as shown in Fig. 5-5 in accordance with ISO 129-1. In addition, the graphical symbol can only be indicated at a position away from the dimension line, except for a cylindrical feature.

● 5.2　Moving average method

Most measurement data of geometrical form, contour profile, and surface texture are a series of data points (series data). A typical example of series data is time-series data, which is a set of values obtained by measuring the temporal changes in a phenomenon and then indexing them in time order. With respect to the form and shape of the workpiece, continuous measurement of the position-dependent variation provides the series data of the geometrical variation. In order to extract useful information from these series data, statistical and data science methods must be applied to the series data. Among these methods, the moving average method is one of the most fundamental and important methods; therefore, this section focuses on the moving average method.

Fig. 5-6 shows a schematic diagram of the simple moving average (SMA) calcu-

Fig. 5-6 は，重み w_j がすべて同じ値の単純移動平均（SMA）の計算方法の概念図である．横軸と縦軸はそれぞれ水平方向の座標 x と高さ方向の座標 z である．Fig. 5-6 において，(x_{i-3}, z_{i-3}) から (x_{i+6}, z_{i+6}) の点が測定で得られた系列データの座標を表し，(x_{i-1}, ZMA_{i-1}) から (x_{i+4}, ZMA_{i+4}) の点が移動平均法により平均化されたデータの座標，w_{j-2} から w_{j+2} の点が単純移動平均の重み関数の値を表している．Fig. 5-6 では重みの点数が 5 点の移動平均であるが，幾何計測分野や計測学分野でも重み関数の点数は奇数個とすることが多い．

単純移動平均の計算は，式 (5-1)，(5-2) のように表すことができる（式では，一般化

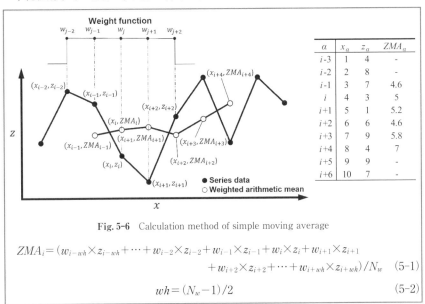

Fig. 5-6 Calculation method of simple moving average

$$ZMA_i = (w_{i-wh} \times z_{i-wh} + \cdots + w_{i-2} \times z_{i-2} + w_{i-1} \times z_{i-1} + w_i \times z_i + w_{i+1} \times z_{i+1}$$
$$+ w_{i+2} \times z_{i+2} + \cdots + w_{i+wh} \times z_{i+wh}) / N_w \quad (5\text{-}1)$$

$$wh = (N_w - 1)/2 \quad (5\text{-}2)$$

lation method in which all values of weights w_j are equal, where the horizontal and vertical axes are the coordinates x and z in the horizontal and vertical directions, respectively. In Fig. 5-6, the points with coordinates from (x_{i-3}, z_{i-3}) to (x_{i+6}, z_{i+6}) represent the series data obtained by measurement, and from (x_{i-1}, ZMA_{i-1}) to (x_{i+4}, ZMA_{i+4}) represent the data averaged by SMA, and the points from w_{j-2} to w_{j+2} represent the values of the weight function of SMA. The weight function of the moving average method in Fig. 5-6 has 5 points; similarly, the number of points in the weight function is almost always an odd number, even in the field of geometrical measurement and metrology.

The calculation of SMA can be expressed as in Eqs. (5-1) and (5-2), where the number of points in the weight function is set to N_w for generalization in these equations. To explain the calculation in detail, as shown in Fig. 5-6, the height coordi-

のために重み関数の点数を N_w と置いている）．具体的には，Fig.5-6 に示すように，高さ方向の座標 z_{i-2} から z_{i+2} のそれぞれに対して重みの値 w_{j-2} から w_{j+2} を掛けて合計し，それらの合計値を重み関数の点数 N_w で割る．同様の計算を測定で得られた系列データに対して逐次的に行うことにより，平均化された値 ZMA_a の系列を得ることができる．Fig.5-6 の右の表は，具体的な数値を用いて単純移動平均を計算した結果である．このような移動平均法を適用することによって，細かな変動成分やノイズなどを取り除いた主要な成分や平滑化されたデータが得られる．

　この移動平均の方法は，幾何計測の分野だけでなく，経済分析や株価分析，気象情報の分析（例えば，気温，気圧）など様々な解析においても利用されている．また，移動平均法の計算アルゴリズムは，畳み込み積分や自己相関関数，相互相関関数，画像用デジタルフィルタ，画像処理，画像認識の計算に応用することができる．

● 5.3　触針先端による不確かさ

接触式の測定機の不確かさの要因の中で重要なものは，触針先端による物理的，幾

nates z_{i-2} to z_{i+2} are multiplied by the weight function values w_{j-2} to w_{j+2}, respectively, and then these values are summed up and then divided by N_w; as a result, an average value ZMA_i is obtained. Performing similar calculations sequentially on all measured series data provides a set of coordinate points: ZMA_a averaged over the measured data. The table on the right side of Fig. 5-6 shows the results of calculating specific numerical values using SMA. The application of these moving average methods, including SMA, provides the smoothed data or the main components of the original signal, with small variation components or noise components being reduced or removed.

This moving average method is used not only in the field of geometrical measurement but also in various analyses, such as economic analysis, stock price analysis, and analysis of weather information (e.g., temperature and atmospheric pressure). Furthermore, such moving average calculation algorithms can be extended to the calculation of convolution, autocorrelation, cross-correlation, digital filter for image, digital image processing, and image recognition.

● 5.3　Uncertainty caused by contact stylus tip

Important factors of uncertainty in contact stylus measurement instruments for surface texture include the effect of the stylus tip, deformation/scratching caused by contact, among others. Among these geometric uncertainties, this section focuses

何学的な影響や接触による変形・きずなどが挙げられる．本節では，触針先端に起因する幾何学的な不確かさのうち，特に谷底への影響について解説する．

　触針先端は有限の半径を持ち無限小にすることはできないため，触針先端球が凹凸の谷底に完全に入り込むことができない．ここで，いくつかの解説書や測定機器メーカの説明資料において Fig. 5-7(a)のような図で触針先端の影響を解説している場合を目にするが，触針先端球を模擬した円は縦横比が1：1で描画されているにもかかわらず Fig. 5-7(a)の粗さ曲線は縦横比が1：1とはなっていないことに注意する必要がある．Fig. 5-7(a)のような，しばしば目にする粗さ曲線は縦横比が1：1/20〜1/250に設定されており，横方向が大幅に縮められている．そのため，触針先端半径の影響を適切に考察するためには，Fig. 5-7(b)の粗さ曲線のように Fig. 5-7(a)の粗さ曲線を水平方向に20倍に引き延ばして縦横比を1：1にする必要がある．

　次に，触針先端半径の深さ方向の影響，すなわち，触針先端が凹凸の谷底に到達で

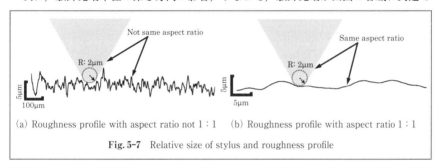

(a) Roughness profile with aspect ratio not 1：1　　(b) Roughness profile with aspect ratio 1：1

Fig. 5-7　Relative size of stylus and roughness profile

on the effect of the stylus tip on the valley bottom.

Since the stylus tip has a finite radius (it cannot be made infinitely small), the sphere of the stylus tip cannot completely reach the valley bottom of the geometrical irregularities. Here, some technical papers or some manuals and brochures from measurement instrument manufacturers explain the effect of the stylus tip using a diagram such as Fig. 5-7(a); it is important to note here that the aspect ratio of the roughness profile shown in Fig. 5-7(a) is not 1:1, although the circle simulating the tip sphere is 1:1. Frequently seen roughness profiles, such as Fig. 5-7(a), have aspect ratios of 1: 1/20 to 1/250, and their horizontal direction is significantly compressed. Therefore, in order to properly consider the effect of the stylus tip radius, the roughness profile in Fig. 5-7(a) need to be stretched horizontally by a factor of 20 to make the aspect ratio 1:1, as shown in the roughness profile of Fig. 5-7(b).

Next, in order to obtain the effect of the stylus tip radius in the depth direction, i.e., the amount by which the stylus tip cannot reach the valley bottom of the geometrical irregularities: Ed, a geometric consideration is made, as shown in Fig. 5-8.

きない量 *Ed* を求めるために，Fig.5-8 のように幾何学的に考察する．Fig.5-8 の幾何学的関係から，触針先端が谷底に到達できない深さ *Ed* は式(5-3) で表すことができる．一般的な機械加工による表面の実際の凹凸の傾斜の角度は，平均値として数°〜15° 未満であることが多く，表面粗さが小さくなるほど角度は小さくなる傾向となる．よって，触針先端半径 r_{tip} を JIS/ISO 規格で標準とされている 2 μm とし，凹凸の傾斜の角度 θ_p を 5° として計算すると，式(5-3)から *Ed* は 0.007 64 μm（7.640 nm）と求められる．

規格で標準とされている触針先端半径 r_{tip} は 2 μm であるが，市販されている最小の触針先端半径は 0.1 μm 以下のものもある．また，一般的な光学式の表面性状測定機の光のスポットの半径は，第9章の光の回折限界に関係する式(9-5)から光波長 $\lambda=$ 0.445 μm，開口数 NA=0.95（9.2 節参照），および，$\lambda=0.633$ μm，NA=0.3 の組み合わせにより FWHM で考えると 0.12〜0.54 μm 程度，$1/e^2$ の場合は 0.19〜0.87 μm

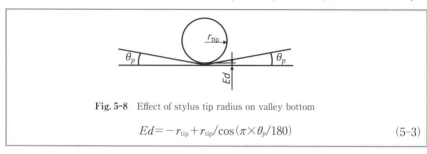

Fig. 5-8　Effect of stylus tip radius on valley bottom

$$Ed=-r_{tip}+r_{tip}/\cos(\pi\times\theta_p/180) \tag{5-3}$$

From Fig.5-8, the depth at which the stylus tip cannot reach the valley bottom: *Ed* is given by Eq. (5-3); where θ_p is the angle of slope (or local gradient) of the surface irregularities, r_{tip} is the stylus tip radius. The slope angle of the actual surface irregularities obtained by widely used machining is often from several degrees to less than 15° ($\pi/12$ rad) as an average angle value, and the angle tends to decrease as the surface roughness decreases. Hence, when r_{tip} is set to 2 μm, which is the default value in the JIS/ISO standard, and θ_p is set to 5° ($\pi/36$ rad), *Ed* is obtained as 0.007 64 μm (7.640 nm) from Eq. (5-3).

The stylus tip radius r_{tip}: 2 μm is the default value in the JIS/ISO standard, on the other hand, some commercially available stylus tips have a minimum radius of 0.1 μm or less. In addition, it should be noted that the radius of the focal spot of general optical surface texture measurement instruments ranges approximately from 0.12 to 0.54 μm when considered in the FWHM with the combination of $\lambda=$ 0.445 μm, NA=0.95 and $\lambda=0.633$ μm, NA=0.3, based on Eq. (9-5), which is related to the diffraction limit of light in Chapter 9 (approximately from 0.19 to 0.87 μm when considered in the $1/e^2$), this is the physical minimum limit of the collection

程度が集光の物理的な限界であることに留意する必要がある．他方，現実的ではないが，接触式の測定機の触針先端半径は原子１つが物理的な限界といえる．以上より，nm からサブ μm レベルの微細な形状を対象とする場合には，触針先端半径や光のスポット径だけでなく，触針先端の頂角や光が対象物に入射する角度に関係する NA にも留意して計測する必要がある．

【演習問題】

5-1）幾何公差である真円度の定義について説明せよ．

5-2）重みがすべて１の矩形重み関数の移動平均（SMA）のプログラムコードを作成せよ．解答するプログラムコードは，計算アルゴリズムのコア部のみでもよく，ヘッダーなどは省略しても構わない．

5-3）Fig. 5-8 の傾斜の角度 θ_p が 3° もしくは 15° となるような谷底を，触針先端半径 r_{tip} が 0.1 μm の触針で測定したとき，触針先端が谷底に到達できない深さ *Ed* を求めよ．

optics. In contrast, although not realistic, the physical limit of the radius of the tip of the stylus of a contact-type measuring machine can be considered as one atom. From the above, when measuring a microscopic structure from nm to sub-μm level, it is essential to pay attention not only the stylus tip radius or the spot size of the light but also to the vertex angle of the stylus tip or NA, which is related to the angle of incidence.

【Problems】

5-1）Explain the definition of the roundness, which is a geometrical tolerance.

5-2）Write a program code for the moving average method using a rectangular weight function (SMA) with weights all equal to 1. Here, the program code to answer can only be the core part of the calculation algorithm, and the header, etc. can be omitted.

5-3）When measuring the bottom of a valley where the slope angle θ_p in Fig. 5-8 is 3° ($\pi/60$) or 15° ($\pi/12$) using a stylus with a tip radius r_{tip} 0.1 μm, calculate the depth, *Ed*, at which the stylus tip cannot reach the valley bottom.

第6章 光干渉計

長さが光速で定義されていることから，光の波長や周波数を基準として計測する光干渉計は，ナノメートルスケールの超精密な "ものさし" として，距離や形状およびそれらの変化（変位）といった幾何学量を代表に様々な分野の精密計測において，よく用いられる計測原理である．その歴史も長く，種々の測定対象に対して，多くの干渉計測手法が提案されている．本章では，光干渉計の基礎原理に触れ，変位，距離，平面形状を測定対象とした代表的な干渉計測法について説明する．

● 6.1 光干渉計の基礎原理

光干渉計では，光源から出た光を分割し，一方を信号光，他方を参照光として用い，それら2つの波の干渉現象に基づいている．まずは，その物理現象について考える．

$$I = A_1^2 + A_2^2 + 2A_1A_2\cos(\varphi_1 - \varphi_2) \quad (6\text{-}1) \qquad \varphi_1 - \varphi_2 = \frac{2\pi}{\lambda}\mathrm{OPD} \quad (6\text{-}2)$$

Chapter 6　Interferometry

Interferometry is a technique to measure geometrical quantities, such as distance, shape, and displacement, based on the wavelength or frequency of light. As length is defined using the speed of light, interferometry is one of the most commonly used measurement principles in precision measurement as an ultra-precise "measurement scale." The history of interferometers is long, and many types of interferometry have been proposed for various measurement targets. This chapter explains the fundamental principle of interferometry, and standard interferometric methods for measuring displacement, distance, and flat surface are described.

● 6.1　Basics of interferometry

In interferometry, the light emitted from a light source is split, and one is used as a signal and the other as a reference. The interferometer is based on the interference phenomenon of these two waves. Assume the two waves have the same frequency and different amplitudes (A_1, A_2). Consider the observation of their interfer-

同じ振動数を持ち，振幅がそれぞれ A_1, A_2 の二つの波を仮定し，それらの干渉波をある位置で観測した場合，干渉波の強度は式(6-1)によって表される．添え字は波1と波2を，φ は波の位相を意味し，$\varphi_1-\varphi_2$ は位相差である．位相差の変化に応じて，干渉波の強度が cos 波状に変化することがわかる．特に光波の場合，光波長を λ とすれば，式(6-2)に示すように，位相差は光路長差（OPD）の変化量に応じ，周期的な干渉波強度の変調（干渉縞）が観測できる．この干渉縞を用いて，様々な変化を測定できる．

　干渉縞の形成には，コヒーレンス（可干渉性）が重要となる．コヒーレンスは時間コヒーレンスと空間コヒーレンスに大別できる．時間コヒーレンスは，光の周波数幅 Δv の逆数で表される（式(6-3)）．コヒーレンス長 L は，コヒーレンス時間 t_c と光速 c を用いて式(6-4)のようになる．Fig. 6-1 に光路長差の変化による干渉縞を示すが，光源の単色性がよく（＝Δv が小さい）時間コヒーレンスがよい場合，光路長差が大き

$$t_c \approx \frac{1}{\Delta v} \quad (6\text{-}3) \qquad\qquad L = ct_c \quad (6\text{-}4)$$

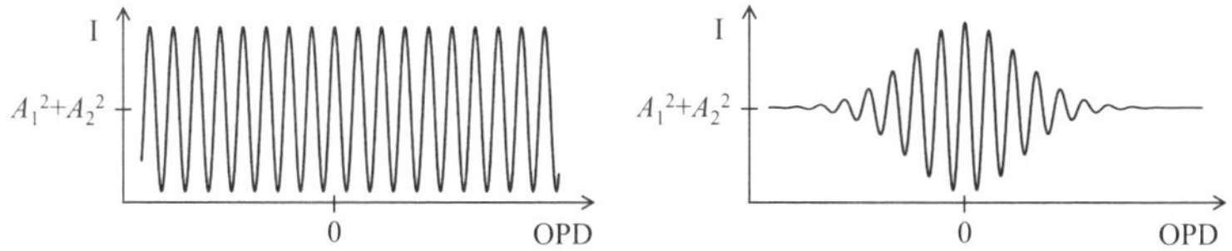

(a) High temporal coherence

(b) Low temporal coherence

Fig. 6-1   Interference fringe with different temporal coherence

ence at a certain position; the intensity of the interference wave can be expressed by Eq. (6-1). The subscripts denote waves 1 and 2, φ means the phase of the waves, and $\varphi_1-\varphi_2$ is the phase difference. The intensity of the interference wave varies according to the cosine function as the phase difference changes. The optical phase difference depends on the amount of change in optical path difference (OPD), and periodic modulation of interference wave intensity (interference fringes) can be observed, as shown in Eq. (6-2), where λ is the wavelength. Various changes in the quantities can be measured based on these interference fringes.

Coherence, divided into temporal and spatial, is an important factor required for forming interference fringes. Temporal coherence is expressed by the inverse of the spectral linewidth Δv (Eq. (6-3)). The coherence length L is expressed by the coherence time t_c and the speed of light c, as in Eq. (6-4). Fig. 6-1 shows interference fringes as a function of the optical path length difference. If the light source is monochromatic (＝Δv is small) and has good temporal coherence, interference fringes are obtained even with a large optical path length difference (Fig. 6-1(a)).

くても干渉縞が得られる（Fig. 6-1(a)）．一方で，波長幅が大きく時間コヒーレンスが悪い場合，2光波の光路長差が0の付近でのみ干渉縞が観測できる（Fig. 6-1(b)）．このように，光路長差がコヒーレンス長を上回ると，干渉縞観察は困難となる．

　空間コヒーレンスは光源の空間的な広がりに起因する．よく光学理論では理想的な点から放出される光源（点光源）を考えるが，現実的には光源は有限径を持つ．光源サイズが大きくなると空間コヒーレンスは低下し，干渉縞のコントラストが下がる（Fig. 6-2）．ただし，干渉縞コントラストはOPDには依存しない．空間コヒーレンスを向上させるためには，光源から出る光をピンホールやシングルモード光ファイバーのような空間フィルタを通す工夫がなされる．

● 6.2　変位計測

ホモダイン法　光干渉計による変位計測として，Fig. 6-3に示すような単一波長光源を用いたホモダイン法によるマイケルソン干渉計を考える．コヒーレンス長を確保す

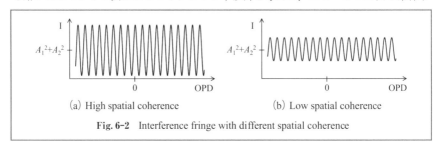

(a) High spatial coherence　　　　(b) Low spatial coherence

Fig. 6-2　Interference fringe with different spatial coherence

Conversely, when the wavelength width is large and the temporal coherence is poor, interference fringes can be observed only near zero optical path length difference (Fig. 6-1(b)). Thus, interference fringes are difficult to observe when the optical path length difference exceeds the coherence length.

　Spatial coherence is attributed to the spatial extent of the light source. A light source is often considered to be emitted from an ideal point source in the theoretical description, but practically, a light source has a finite size. The spatial coherence of light is degraded as the light source size increases, resulting in lower contrast in the interference fringes (Fig. 6-2). However, the contrast is independent of the OPD. Spatial filters, such as a pinhole or a single-mode optical fiber, are used to improve spatial coherence.

● 6.2　Displacement measurement

Homodyne detection　As a displacement measurement using interferometry, consider a Michelson interferometer based on homodyne detection using a monochro-

るため，光源にはレーザが用いられる．レーザから出た光は，ビームスプリッタで2分割され，それぞれ参照面と測定面で反射し，再度ビームスプリッタを経て同一光路となり干渉する．それぞれの光は式(6-5), (6-6)のように表せる．ただし，Eは電界，Aは振幅，ωは角周波数，tは時間，kは波数，xは距離，φは初期位相である．光検出器は干渉した光を検出するため，E_1とE_2の重ね合わせを考える（式(6-7)）．光波の振動は非常に高速（例えば，波長633 nmのときは約470 THz（テラヘルツ，テラは10^{12}））であり，現実的には光検出器の出力は時間平均強度となる．式(6-8)右辺の第1項，第2項は参照光と測定光それぞれの時間平均強度を意味し，第3項が干渉項である．〈 〉は時間平均を意味する．簡単のため，測定光と参照光の振幅は等しい

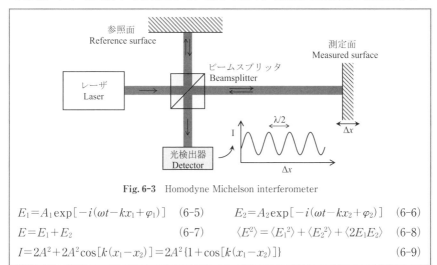

Fig. 6-3 Homodyne Michelson interferometer

$$E_1 = A_1 \exp[-i(\omega t - k x_1 + \varphi_1)] \quad (6\text{-}5) \qquad E_2 = A_2 \exp[-i(\omega t - k x_2 + \varphi_2)] \quad (6\text{-}6)$$

$$E = E_1 + E_2 \quad (6\text{-}7) \qquad \langle E^2 \rangle = \langle E_1^2 \rangle + \langle E_2^2 \rangle + \langle 2E_1E_2 \rangle \quad (6\text{-}8)$$

$$I = 2A^2 + 2A^2 \cos[k(x_1 - x_2)] = 2A^2\{1 + \cos[k(x_1 - x_2)]\} \quad (6\text{-}9)$$

matic light source, as shown in Fig. 6-3. A laser is used as the light source to ensure coherence length. The laser beam is divided into two beams by a beamsplitter, reflected at the reference and measured surfaces, respectively, and then passes through the beamsplitter again to align the same light path and interfere with each other. Each light can be expressed as in Eqs. (6-5) and (6-6), where E is electric field, A is amplitude, ω is angular frequency, t is time, k is wavenumber, x is distance, and φ is the initial phase. As the detector detects the interfered beam, consider the interference of E_1 and E_2 (Eq. (6-7)). The frequency of light is very high (e.g., about 470 THz (Tera, T, means 10^{12}) at a wavelength of 633 nm). Therefore, the output of the detector is the time-averaged intensity. The first and second terms on the right side of Eq. (6-8) refer to the time-averaged intensity of the reference and measurement beams, respectively, and the third term is the interference term. 〈 〉

$(A_1=A_2=A)$ と考え，また初期位相を 0 $(\varphi_1=\varphi_2=0)$ と置き，式(6-8)を整理すると，式(6-9)が得られる．これからわかるように，干渉光強度は参照面と測定面の光路長差 x_1-x_2 に依存する．参照面は固定しており，測定面が変位するが，測定面変位量に応じた干渉光の強度変調が起こり，測定された光強度の変化から測定面の変位を測定する．測定面が Δx 変位するとき，光路長差は往復で $2\Delta x$ 変化することに注意すると，光強度は測定面変位に対して周期 $2\Delta x/\lambda$ で変化する．つまり，変位 $\lambda/2$ で位相 2π の変化となる（波数 $k=2\pi/\lambda$）．光波長 632.8 nm を用いた場合，変位 316.4 nm で 1 周期の干渉縞が得られる．ナノスケール分解能計測のために，位相計測により干渉縞を内挿する．位相測定法については，4 位相法（7 章を参照）がよく知られているが，変位計測ではその場の位相計測が必要であるため，90° ずつ位相のずれた 4 つの干渉信号を位相素子や偏光素子を用いて生成・計測し，処理することで干渉信号の位相を得る．

ヘテロダイン法 ヘテロダイン干渉計は，わずかに異なる 2 つの周波数を持つ光源を

means time-averaged. For simplicity, considering that the amplitude of the measurement and the reference beam are equal $(A_1=A_2=A)$ and the initial phase is 0 $(\varphi_1=\varphi_2=0)$. Thus, Eq. (6-9) can be derived from Eq. (6-8). As can be seen, the interferometric intensity depends on the optical path length difference between the reference and measured surfaces, x_1-x_2. Since the reference surface is fixed, the displacement of the measured surface causes intensity modulation of the interference beam. Therefore, the displacement of the measured surface is evaluated from the measured light intensity of the interference beam. Note that when the measured surface is displaced by Δx, the optical path length difference changes by $2\Delta x$ for a round trip, and the light intensity changes periodically with period $2\Delta x/\lambda$ with respect to the measured surface displacement; that is, a phase change of 2π at displacement $\lambda/2$ $(k=2\pi/\lambda)$. When a wavelength of 632.8 nm is used, one period of interference fringes is obtained at a displacement of 316.4 nm. Interpolating interference fringes using phase measurement enables the measurement of displacement with a resolution in nanometric scales. The four-step method (see Chapter 7) is a well-known phase-measurement method. As real-time phase measurement is required for displacement measurement, four interference signals with a different phase by 90° are generated and measured using phase and polarization optical elements. The phase of the interference light is obtained by processing them.

Heterodyne detection A heterodyne interferometer uses a light source with two

用いた光干渉計である．この2つの周波数の差をビート周波数と呼ぶ．ビート信号の周波数変化や位相変化から物体の変位を計測する．Fig. 6-4 にその構成を示す．光源として，異なる周波数 ω_1, ω_2 を持ち，それぞれ直交した2つの偏光を持つ光を用いる．まず，ビームスプリッタで2分割された一部の光が偏光ビームスプリッタに向かう．周波数 ω_2 の光は偏光ビームスプリッタで光路を転換し，周波数 ω_1 の光は測定対象で反射して偏光ビームスプリッタに再入射する．透過軸を 45° に傾けた偏光板を通り，光検出器でその干渉信号を測定する．いま，わずかに周波数の異なる光の干渉を考える．それぞれの複素振幅は式(6-10), (6-11)のようになり，光検出器2から出力さ

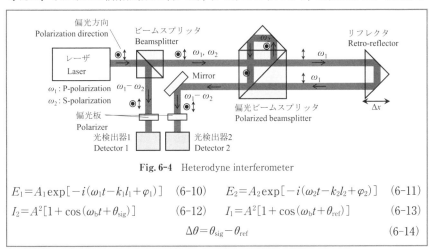

Fig. 6-4　Heterodyne interferometer

$$E_1 = A_1 \exp[-i(\omega_1 t - k_1 l_1 + \varphi_1)] \quad (6\text{-}10) \qquad E_2 = A_2 \exp[-i(\omega_2 t - k_2 l_2 + \varphi_2)] \quad (6\text{-}11)$$

$$I_2 = A^2[1 + \cos(\omega_b t + \theta_{sig})] \quad (6\text{-}12) \qquad I_1 = A^2[1 + \cos(\omega_b t + \theta_{ref})] \quad (6\text{-}13)$$

$$\Delta\theta = \theta_{sig} - \theta_{ref} \quad (6\text{-}14)$$

slightly different frequencies. The difference between these two frequencies is called the beat frequency, and the displacement of an object is measured from the frequency and phase changes of the beat signal. Fig. 6-4 shows the configuration of the measurement system. Two orthogonal polarizations of light are used as the light source. Each polarization component has a slightly different frequency, ω_1 and ω_2. First, the beam divided into two parts by the beamsplitter, which are directed to the polarizing beamsplitter. The beam with frequency ω_2 is reversed by the polarizing beamsplitter, and the beam with frequency ω_1 is reflected by the retroreflector (the measured target) and re-enters the polarizing beamsplitter. The beams pass through the polarizer with the transmission axis tilted at 45°. The detectors measure the interference signal. Here, let us consider the interference of beams with slightly different frequencies. The complex amplitudes are as in Eqs. (6-10) and (6-11), and the output signal I_2 from Detector 2 can be expressed as in Eq. (6-12), where ω_b is the frequency difference between ω_1 and ω_2, and θ_{sig} is $k\Delta l$

れる信号 I_2 は式（6-12）のように表すことができる．ただし，ω_b は ω_1 と ω_2 の周波数差であり，θ_{sig} は $k\Delta l + (\varphi_1 - \varphi_2)$ である（$\Delta l = l_1 - l_2$）．時間と共に干渉信号が周波数差 ω_b の周期で変化することがわかる．「ホモダイン法」で述べたように光周波数は 100 THz オーダであり光検出器では時間平均強度しか測定できないが，ビート周波数が数十MHzとなるように設定することで干渉光の時間変化が測定可能となる．リフレクタが Δx 変位すると信号光の光路長が変化する．その際の信号光位相 θ_{sig} を求めるため，光検出器1から出力される信号 I_1 に着目する．その出力は式(6-13)のように表される．ここで，θ_{ref} は $(\varphi_1 - \varphi_2)$ である。この θ_{sig} と θ_{ref} の位相差 $\Delta\theta$（式(6-14)）を位相計を用いて電気的に計測する．θ_{ref} は変化しないため，$\Delta\theta$ より Δl を求められる．

● 6.3　距離計測

合致法　光干渉による変位計測は高精度であるが，距離計測として用いるには，測定距離が $\lambda/2$ 程度と非常に短く適用が困難である．しかし，測定物のおよその距離が既知の場合，複数波長を用いた合致法により距離測定ができ，ブロックゲージの校正などに用いられている．Fig. 6-5(a)のように，マイケルソン干渉計を想定し，ブロック

$+ (\varphi_1 - \varphi_2)$, where $\Delta l = l_1 - l_2$. The interference signal modulates with the period of the frequency difference ω_b. As described in "Homodyne detection", the frequency of visible light is of the order of 100 THz, so the time-averaged intensity is the output of a detector. However, for the beat frequency, which is of the order of 10 MHz, the time variation of the interference beam can be measured. When the retro-reflector is displaced by Δx, the optical path length of the signal beam changes. To obtain the phase θ_{sig} of the signal beam, the output signal I_1 from Detector 2 is considered. The output is expressed as in Eq. (6-13), where θ_{ref} is $(\varphi_1 - \varphi_2)$. The phase difference $\Delta\theta$ between θ_{sig} and θ_{ref} (Eq. (6-14)) is measured electrically using such as a phase meter. Δl is obtained from $\Delta\theta$, since θ_{ref} is fixed.

● 6.3　Distance measurement

Coincidence method　Interferometric displacement measurement boasts high accuracy, but the measurable distance is limited to approximately $\lambda/2$, rendering it suitable only for distance measurement within this range. However, if the approximate length of the measured object is known, interferometry applies to distance measurement, in which the coincidence method with multiple wavelengths is used. This method is used to calibrate gauge blocks. As shown in Fig. 6-5(a), supposing a Mi-

ゲージは平面度の高い平面基板上に設置している．参照面をやや傾けると，CCD では
ブロックゲージの上端および基板からの反射光に対する干渉縞が得られる（Fig. 6-5
(b)）．上面と基板面の差を考えると，ブロックゲージ高さは式(6-15)となる．L はブ
ロックゲージの高さ，λ は光波長，m は干渉縞次数，e は干渉縞からの端数である
（Fig. 6-5(c)）．e は測定可能であるが，m が未知である．この m を求めるため，3 つ
の波長を用いる．まず波長 λ_1 に対し，ある程度既知な長さ La から推測される m_1 を
得る．これと測定値 e_1 を用いて，波長 λ_1 による推定距離 L_1 を算出する．L_1 を用いて，
式(6-15)から波長 λ_2, λ_3，推定量 m_2, m_3 を用いた時の端数 e_2', e_3' を算出する．この

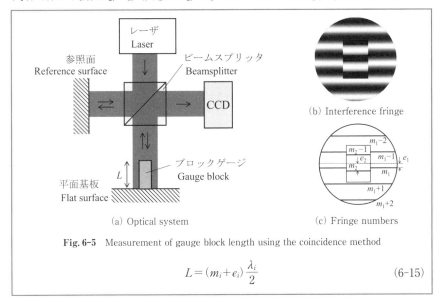

(a) Optical system   (c) Fringe numbers

Fig. 6-5 Measurement of gauge block length using the coincidence method

$$L = (m_i + e_i)\frac{\lambda_i}{2} \tag{6-15}$$

chelson interferometer, the gauge block is installed on a flat substrate with high
flatness. When the reference surface is slightly tilted, interference fringes are obtained for the light reflected from the top surface of the gauge block and the flat
surface (Fig. 6-5(b)). Given the difference between these two surfaces, the length
of the gauge block can be expressed as Eq. (6-15), where L is the length of the
gauge block, λ is the wavelength, m is the interference fringe order, and e is the
fraction from the interference fringe (Fig. 6-5(c)). e can be experimentally measured, and m is unknown. In order to estimate m, three wavelengths are employed.
First, m_1 is estimated for wavelength λ_1, using the a priori known estimated length
L_a. Using m_1 and the measured value e_1, the length L_1 is calculated. Then, fractions
e_2' and e_3' with wavelengths λ_2 and λ_3 are calculated using L_1, m_2, and m_3 from
Eq. (6-15). m_2 and m_3 should be obtained in the same manner as for m_1. When these

e_2', e_3' が測定値 e_2, e_3 を合致するとき，これを測定結果とする．一致しない場合，e' と e が合致するよう，干渉縞次数 m を $m \pm 1$，$m \pm 2$ などに置き換えて探索する．得られた m と e から L を求める．

分光干渉計　分光干渉計は，広い波長幅を持った光源を用い，波長ごとの干渉強度を測定することで距離計測を行う．Fig. 6-6(a)に分光干渉計の構成例を示す．光源は LED 光源を考える．また，光の検出器には，各波長の強度を測定できる分光器を用いる．光は，参照面と距離 L だけ離れた測定面で反射し，それぞれの反射2光波が干渉するが，$N_i \times \lambda_i$（N は整数）が $2L$ となる波長 λ_i で光は強め合う（Fig. 6-6(b)）．測定

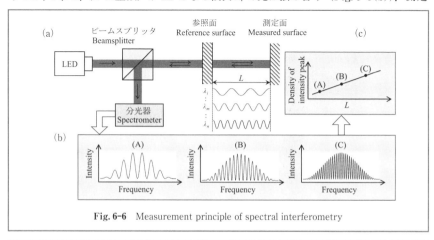

Fig. 6-6　Measurement principle of spectral interferometry

e_2' and e_3' coincide with the measured values e_2 and e_3, the measured length is calculated using these values. If not coincident, the interference fringe order m is replaced with $m \pm 1$, $m \pm 2$, etc., to search for a match between e' and e. Finally, L is valuated from the obtained m and e.

Spectral interferometry　This method uses a light source with a wide wavelength range. Distance can be measured by analyzing the interference intensity at each wavelength. Fig. 6-6(a) shows an example configuration of a spectral interferometer. An LED light source with a wide wavelength range is used as the light source. A spectrometer that can measure the intensity of light at each wavelength is used as a light detector. The laser beam is reflected at a reference surface and a measured surface. Their separated distance is L. Each reflected light wave interferes with each other, and the light intensifies at wavelength λ_i, where $N_i \times \lambda_i$ (N is an integer) is $2L$ (Fig. 6-6(b)). The constructive interference wavelength becomes

面の距離が大きくなると，強め合う波長の間隔は密になる．得られた干渉スペクトル
の周波数を軸に周波数解析する．その周波数と測定距離 L が対応するため，測定面距
離を測定することができる（Fig. 6-6(c)）．

● 6.4 平面計測

フィゾー干渉計 フィゾー干渉計は，平面に近い測定面の凹凸を数 nm の精度で計測
する手法である．平面度の非常に高い基準面（オプティカルフラット）を参照面とし
て測定物に向かい合わせ，参照面と測定物からの反射光を干渉させる（Fig. 6-7）．平
面凹凸を計測するため，検出器は CCD などのイメージセンサを用い，光源にはレー

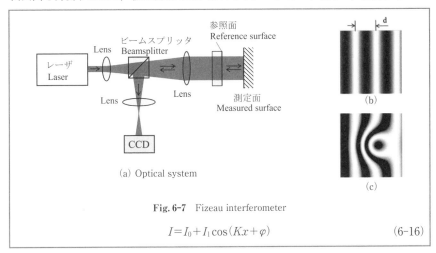

Fig. 6-7 Fizeau interferometer

$$I = I_0 + I_1 \cos(Kx + \varphi) \tag{6-16}$$

dense as the distance between the measured and reference surfaces increases. Frequency analysis is performed on the frequency of the obtained interference spectrum. As the frequency corresponds to the measurement distance L, the measurement surface distance can be measured (Fig. 6-6(c)).

● 6.4 Flat surface measurement

Fizeau interferometry This technique involves measuring the surface shape of a
plane surface and achieving an accuracy of a few nm. A reference plane, such as an
optical flat, is placed in front of the measured surface as a reference surface. The
reflected light from the reference and the measured surfaces interfere, as shown in
Fig. 6-7. An image sensor such as a CCD is used as the detector to measure surface
shape, and a monochromatic laser is used as the light source. The phase of the interfered light changes in response to variations in the optical path length at the

ザを用いる．参照面と測定面の光路長の差に応じて干渉光の位相が変化する．測定物
の面内方向の位置による干渉光強度の違い（干渉縞）を観察するが，その干渉縞は式
(6-16)で表される．I_0 は干渉縞の平均強度，I_1 は干渉縞の振幅，φ は基準面と測定物
からの反射光の位相差，K は干渉縞の波数（干渉縞間隔を d とすると，$K=2\pi/d$），x
は面内方向の位置を表す．干渉縞波数 K は参照面と測定面の交差角に依存する．参照
面と測定面が理想的な平面である場合は Fig. 6-7(b) のような一方向の干渉縞が得ら
れ，測定面に凹凸があると Fig. 6-7(c) のように干渉縞が歪む．歪みから測定面の表
面形状を計測可能である．参照面を基準とした測定面の光軸方向のずれは $\delta=(\lambda/4\pi)\varphi$
となる（位相計測方法は，第7章を参照）．第12章で述べるように，測定結果には参
照面の形状も含まれ，この参照面の平面度が測定不確かさの主要因となる．

reference and measurement surfaces. The interference light intensity distribution,
i.e., the intensity distribution of the interference fringes, is expressed by Eq. (6-16).
I_0 is the averaged light intensity, I_1 is the amplitude of the interference fringe, φ is
the phase difference of the reflected light from the reference and measured surfaces, K is the wavenumber of the interference fringes ($K=2\pi/d$, d is the interference
fringe spacing), and x is the in-plane position. K relies on the angle formed between
the reference and measured surfaces. If both the reference and measured surfaces
are ideal planes, interference fringes in one direction are obtained, as shown in
Fig. 6-7(b). The interference fringes are distorted depending on the shape of the
measured surface, as shown in Fig. 6-7(c). The surface shape can be estimated
from distorted interference fringes. The deviation of the measured surface from the
reference surface is $\delta=(\lambda/4\pi)\varphi$ (see Chapter 7 for the phase-measurement method). Note that the measurement results also include the shape of the reference surface, which is the main cause of the uncertainty of measurement in the flat surface
measurement.

【演習問題】

6-1) 中心波長 633 nm である光源を考え，波長線幅が 1 pm と 1 nm であるときのコヒーレンス長をそれぞれ計算せよ．

6-2) 式 (6-5) と (6-6) から式 (6-9) を導出せよ．ただし，$A = A_1 = A_2$ とする．

6-3) 公称値 $L = 200$ mm のブロックゲージを合致法で校正する場合を考える．3 つの波長は $\lambda_1 = 633$ nm, $\lambda_2 = 532$ nm, $\lambda_3 = 488$ nm とし，干渉縞からの端数がそれぞれ，$e_1 = 0.89$, $e_2 = 0.12$, $e_3 = 0.59$ であった場合，ブロックゲージの校正値はいくらになるか求めよ．

6-4) フィゾー干渉計では，参照面と測定面を完全な平行とはせずに，やや傾けて干渉縞を形成させる．その理由について考察せよ．

【Problems】

6-1) Considering a light source with a center wavelength of 633 nm, calculate the coherence length when the wavelength linewidth is 1 pm and 1 nm.

6-2) Derive Eq. (6-9) from Eqs. (6-5) and (6-6) by assuming $A = A_1 = A_2$.

6-3) Consider the case where a gauge block of nominal value $L = 200$ mm is calibrated by the coincidence method. The three wavelengths used are $\lambda_1 = 633$ nm, $\lambda_2 = 532$ nm, and $\lambda_3 = 488$ nm, and the fractions from the interference fringe are $e_1 = 0.89$, $e_2 = 0.12$, and $e_3 = 0.59$, respectively. Calculate the calibrated value of the gauge block.

6-4) In a Fizeau interferometer, the reference and measured surfaces are not aligned perfectly parallel but slightly tilted to form interference fringes. Discuss the reason for the slight tilt.

第7章　マシンビジョン

　マシンビジョンは画像から物体の位置，形状などの情報を抽出して利用する技術を広く意味する．マシンビジョンは非接触の測定が可能，2次元・3次元の情報が1台もしくは2台のカメラで得られる，測定範囲の自由度が高い，などの利点から位置決め装置や検査装置などでの応用例が多い．また，波長の選択や画像処理の方法によって，位置，形状，温度，振動など様々な対象の計測が可能である．本章では，画像から物体の位置や形状を計測する基本的な方法について述べる．

● 7.1　透視投影モデル

　Fig. 7-1 に示すように，計測対象物をカメラで撮影して，撮影された画像から対象物上の点 A の位置を求めることを考える．ここでは3つの座標系，すなわち①空間に固定されたワールド座標系，②カメラのレンズ中心を原点として光軸を Z 軸に持つカ

Chapter 7　Machine Vision

　Machine vision is a technology for extracting and utilizing information such as the position and shape of an object from an image. Machine vision has been widely applied to positioning and inspection equipment owing to its advantages, such as noncontact measurement, ability to obtain 2- and 3-dimensional information with one or two cameras, and high degree of freedom in the measurement range. In addition, depending on the choice of wavelength and image processing method, various quantities such as position, shape, temperature, and vibration can be measured. This chapter describes the basic methods for measuring the position and shape of objects from images.

● 7.1　Perspective projection model

　As shown in Fig. 7-1, the object to be measured is captured by a camera, and the position of point A on the object is determined from the captured image. In Fig. 7-1, for simplicity, we consider the case in which the world coordinate system and the

メラ座標系，③撮影された画像における画像座標系がある．Fig. 7-1 では簡単のため
に，ワールド座標系とカメラ座標系が一致した場合を考えている．計測対象上のある
点 A のワールド座標を(X_w, Y_w, Z_w)とし，点 A が画像座標(x, y)である点 A_I として撮
影されたとする．すると，点 A の位置を求めることは，画像座標(x, y)からワールド
座標(X_w, Y_w, Z_w)を求める問題となる．

　本来カメラにおいて，レンズを通して撮影される画像はレンズよりも後ろに結像さ
れる倒立像であるが，ここでは便宜上レンズの前，すなわちレンズと計測対象の間に
結像される正立像としている．こうすると像が反転せず，ワールド座標と画像座標の
対応がわかりやすい利点がある．このようなモデルを透視投影モデルと呼ぶ．

　点 A のワールド座標 X_w, Y_w は幾何光学的な関係から導出された式(7-1), (7-2)を
用いて求めることができる．ここで，f はレンズの焦点距離である．しかし，実際は

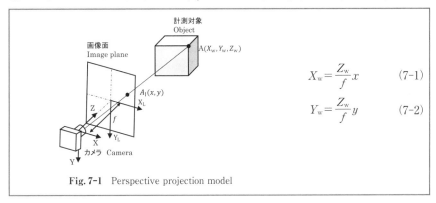

$$X_w = \frac{Z_w}{f} x \qquad (7\text{-}1)$$

$$Y_w = \frac{Z_w}{f} y \qquad (7\text{-}2)$$

Fig. 7-1　Perspective projection model

camera coordinate system coincide. We also assume that the world coordinates of
point A on the measurement target are (X_w, Y_w, Z_w), and point A is captured as
point A_I with image coordinates (x, y). Then, measuring the position of point A is a
problem of finding the world coordinates (X_w, Y_w, Z_w) from the image coordinates
(x, y).

Originally, the image captured through the lens of a camera is an inverted image
formed behind the lens. However, here, for convenience, it is an upright image
formed in front of the lens, that is, between the lens and the object to be measured.
In this way, the image is not inverted, and the correspondence between the world
coordinates and the image coordinates is easy to understand. Such a model is called
a perspective projection model.

The world coordinates X_w and Y_w of point A can be obtained using Eqs. (7-1)
and (7-2), where f is the focal length of the lens. In reality, however, because Z_w is
unknown, the 3D coordinates of the object to be measured cannot be obtained from

Z_w がわからないため，1 台のカメラによる画像のみからでは計測対象の 3 次元座標を求めることはできない．そこで，次節以降で述べる工夫によって計測対象の 3 次元座標または形状を求める．

● 7.2　ステレオ法

マシンビジョンにおいて，3 次元位置を求める場合によく用いられる方法の 1 つにステレオ法がある．ステレオ法では，Fig. 7-2 に示すように，2 台のカメラによって撮影した 2 枚の画像から計測対象の 3 次元位置を求める．まずは単純化のために，左カメラのカメラ座標系とワールド座標系が一致しているとし，右カメラは左カメラを X 方向に平行移動させた位置にあるとする．このようなカメラ配置を平行ステレオと呼ぶ．計測対象上の点 A が左右のカメラで撮影した画像において，点 A_IL と点 A_IR にそ

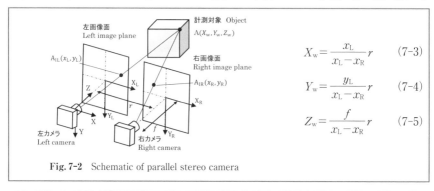

$$X_\mathrm{w} = \frac{x_\mathrm{L}}{x_\mathrm{L} - x_\mathrm{R}} r \qquad (7\text{-}3)$$

$$Y_\mathrm{w} = \frac{y_\mathrm{L}}{x_\mathrm{L} - x_\mathrm{R}} r \qquad (7\text{-}4)$$

$$Z_\mathrm{w} = \frac{f}{x_\mathrm{L} - x_\mathrm{R}} r \qquad (7\text{-}5)$$

Fig. 7-2　Schematic of parallel stereo camera

the image taken by only one camera. Therefore, the 3D coordinates or shape of the object are obtained using the method described in the next section.

● 7.2　Stereo method

The stereo method is widely used in machine vision to obtain a 3D position. In this method, the 3D position of the measurement target is obtained from two images taken by two cameras, as shown in Fig. 7-2. First, for simplicity, the camera coordinate system of the left camera and the world coordinate system are assumed to coincide, with the right camera being positioned with the parallel shift of the left camera in the X direction. Such a camera arrangement is called a parallel stereo. Assuming that point A on the measurement target is projected onto points A_IL and A_IR in the images taken by the left and right cameras, respectively, the world coordinates of point A can be obtained using the image coordinates $(x_\mathrm{L}, y_\mathrm{L})$ and $(x_\mathrm{R}, y_\mathrm{R})$ of point A_IL and point A_IR as in Eqs. (7-3)-(7-5), where r, is the distance between

れぞれ投影されたとすると，点 A のワールド座標は点 A_{IL} と点 A_{IR} の画像座標 (x_L, y_L) と (x_R, y_R) を用いて式(7-3)〜(7-5)のように求められる．ここで，r は左右のカメラ間の距離であり，基線長と呼ぶ．

式(7-3)〜(7-5)の導出のために，Fig.7-3 に示す XZ 平面への投影図を考える．ここで，O_L と O_R はそれぞれ左カメラと右カメラのレンズの中心である．三角形 $AA_{IL}A_{IR}$ と三角形 A O_LO_R が相似であることから，式(7-6)が成り立つ．式(7-6)を整理すると，式(7-5)と同義であることがわかる．また，同様に式(7-7)が成り立つが，式(7-7)に式(7-5)を代入すると式(7-3)が得られる．YZ 平面にて同様の三角形の相似を考えれば，式(7-4)も導出できる．この原理は8.3節で述べる三角測量と同様である．

平行ステレオでは3次元位置を簡単な式で求めることができるが，左右のカメラで

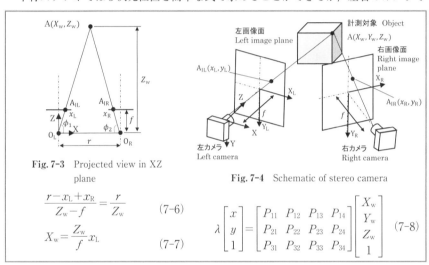

Fig. 7-3 Projected view in XZ plane

Fig. 7-4 Schematic of stereo camera

$$\frac{r - x_L + x_R}{Z_w - f} = \frac{r}{Z_w} \qquad (7\text{-}6)$$

$$X_w = \frac{Z_w}{f} x_L \qquad (7\text{-}7)$$

$$\lambda \begin{bmatrix} x \\ y \\ 1 \end{bmatrix} = \begin{bmatrix} P_{11} & P_{12} & P_{13} & P_{14} \\ P_{21} & P_{22} & P_{23} & P_{24} \\ P_{31} & P_{32} & P_{33} & P_{34} \end{bmatrix} \begin{bmatrix} X_w \\ Y_w \\ Z_w \\ 1 \end{bmatrix} \qquad (7\text{-}8)$$

the left and right cameras, is called the baseline length.

To derive Eqs. (7-3)-(7-5), the projection onto the XZ plane is considered as shown in Fig. 7-3. Here, O_L and O_R are the centers of the lenses of the left and right cameras, respectively. As triangle $AA_{IL}A_{IR}$ and triangle AO_LO_R are similar, Eq. (7-6) holds. Re-arranging Eq. (7-6) shows that it is similar to Eq. (7-5). Similarly, Eq. (7-7) holds. Substituting Eq. (7-5) into Eq. (7-7), we obtain Eq. (7-3). By considering the similarity of the triangles in the YZ plane, we can also derive Eq. (7-4). This principle is similar to that of triangulation, which is described in Section 8.3.

In parallel stereo, the 3D position can be obtained by a simple formula, but the measurement range is narrow because only the common shooting range of the left and right cameras can be measured. Therefore, we consider a more general case

共通する撮影範囲でしか計測できないため，測定範囲が狭い．そこで，より一般的に，Fig. 7-4 に示すように左右のカメラが平行でない場合を考える．一般に，ワールド座標と画像座標の関係は，同次変換行列を用いて式(7-8)で表すことができる．ここで，P_{ij} ($i=1, 2, 3$, $j=1, 2, 3, 4$) はワールド座標におけるカメラの位置と方向，カメラの焦点距離などから決まる係数であり，λ はスケールファクターである．P_{ij} を要素とする行列はカメラ行列と呼ばれている．

左カメラで撮影した画像に関して式(7-8)を展開すると式(7-9)～(7-11)が得られる．ここで，添え字の L は左カメラに関するパラメータであることを意味する．式(7-11)を式(7-9)と(7-10)にそれぞれ代入すると，式(7-12)，(7-13)が得られる．同様にして右カメラで撮影した画像に関して式(7-14)，(7-15)が得られる．ここで，添え字の R は右カメラに関するパラメータであることを意味する．

カメラ行列と画像座標が既知であるとすると，未知数であるワールド座標3つに対して式(7-12)～(7-15)の4つの式が得られるため，連立方程式を解くことでワールド

$$\lambda x_L = P_{L11} X_w + P_{L12} Y_w + P_{L13} Z_w + P_{L14} \tag{7-9}$$

$$\lambda y_L = P_{L21} X_w + P_{L22} Y_w + P_{L23} Z_w + P_{L24} \tag{7-10}$$

$$\lambda = P_{L31} X_w + P_{L32} Y_w + P_{L33} Z_w + P_{L34} \tag{7-11}$$

$$(P_{L11} - P_{L31} x_L) X_w + (P_{L12} - P_{L32} x_L) Y_w + (P_{L13} - P_{L33} x_L) Z_w = P_{L34} x_L - P_{L14} \tag{7-12}$$

$$(P_{L21} - P_{L31} y_L) X_w + (P_{L22} - P_{L32} y_L) Y_w + (P_{L23} - P_{L33} y_L) Z_w = P_{L34} y_L - P_{L24} \tag{7-13}$$

$$(P_{R11} - P_{R31} x_R) X_w + (P_{R12} - P_{R32} x_R) Y_w + (P_{R13} - P_{R33} x_R) Z_w = P_{R34} x_R - P_{R14} \tag{7-14}$$

$$(P_{R21} - P_{R31} y_R) X_w + (P_{R22} - P_{R32} y_R) Y_w + (P_{R23} - P_{R33} y_R) Z_w = P_{R34} y_R - P_{R24} \tag{7-15}$$

where the left and right cameras are not parallel, as shown in Fig. 7-4. In general, the relationship between world coordinates and image coordinates can be expressed by Eq. (7-8), where P_{ij} ($i=1, 2, 3, j=1, 2, 3, 4$) are coefficients determined from the relative position between two cameras, the focal length of the camera, and so on. λ is a scale factor. The matrix with P_{ij} is called the camera matrix.

Expanding Eq. (7-8) for an image taken by the left camera yields Eqs. (7-9)-(7-11), where the subscript L indicates that the parameter is related to the left camera. Substituting Eq. (7-11) into Eqs. (7-9) and (7-10), Eqs. (7-12) and (7-13) are obtained, respectively. Similarly, Eqs. (7-14) and (7-15) are obtained for the image taken by the right camera where the subscript R denotes that the parameter is related to the right camera.

When the camera matrix and image coordinates are known, four Eqs. (7-12)-(7-15) can be obtained for the three unknown world coordinates. Thus, the world coordinates can be obtained by solving simultaneous equations. In practical use, the

座標を求めることができる．実用では最小二乗法を用いて解くことが多く，3台以上
のカメラを用いた場合でも同様に最小二乗解を得ることができる．

　ステレオ法では，同じワールド座標の点が，左右のカメラの画像において異なる画
像座標に撮影されることを利用している．つまり，ステレオ法を適用するためには，
左右のカメラの画像において，対応する点のペアを探す必要がある．この作業をステ
レオマッチングと呼ぶ．Fig. 7-2 や Fig. 7-4 に示す直方体の頂点のようなわかりやす
い特徴点がある場合と異なり，模様のない対象物やなだらかな曲面などは対応する点
の探索が難しい．

　そこで，Fig. 7-5 に示すように，構造化光（幾何学的なパターンの光）を対象物に
投影することで，この対応付けを容易にする工夫がある．プロジェクタを用いた投影
パターンは既知とできるので，カメラで撮影した画像の中でパターンが映っている部

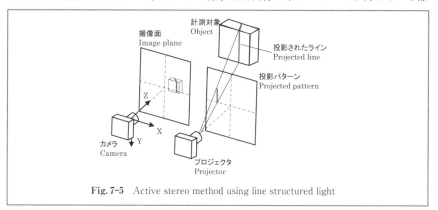

Fig. 7-5 Active stereo method using line structured light

least-squares method is often used to solve the equations, and the least-squares
solution can be obtained in the same way when three or more cameras are used.

　The stereo method is based on the fact that points with the same world coordinates are captured at different image coordinates in the left and right camera images. In other words, to apply the stereo method, a pair of corresponding points in the
left and right camera images must be determined. In contrast to cases where there
are obvious feature points, such as the vertices of the rectangular objects shown in
Fig. 7-2 and Fig. 7-4, determining corresponding points on objects without patterns
or with gently curved surfaces is difficult.

　As shown in Fig. 7-5, a technique exists to facilitate this correspondence by pro-
jecting structured light (light with a geometric pattern) onto the object. Because
the projection pattern is ascertainable, searching for pairs of corresponding points
by focusing only on the captured patterns in the image is sufficient. This method is
called the active stereo method because it actively creates feature points for corre-

分にのみ注目して対応する点のペアを探索すればよい．能動的に対応付けのための特
徴点を作り出しているという意味で，このような方法を能動的ステレオ法と呼ぶ．こ
れに対して，Fig. 7-2 や Fig. 7-4 に示す方法は受動的ステレオ法と呼ぶ．Fig. 7-5 に示
すラインパターンを用いた能動的ステレオ法は光切断法と呼ばれることもあり，広く
産業界で応用されている．この他にも，2 次元的な格子パターンやランダムドットパ
ターンなど，様々な構造化光による計測方法が提案・実用化されている．

　構造化光に縞パターンを用いた場合は，位相シフト法を用いてパターンの位置を高
精度に求めることができる．位相シフト法では，Fig. 7-6 に示すように，正弦波状の
輝度パターンを持つ縞を用いる．このとき，ある計測対象点での輝度は式(7-16)で表
される．ここで，I は輝度変化の振幅，は位相，I_{off} は輝度のオフセットである．パター
ン上の位置は位相に対応する．そこで，投影する縞の位相を意図的に変化させて（シ
フトさせて）複数回撮影し，計測結果を組み合わせることで位相を求める．

　Fig. 7-6 には $\pi/2$ rad ずつ位相をシフトさせた 4 つの縞パターンを示している．こ

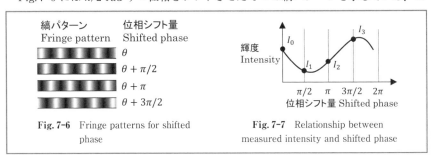

| Fig. 7-6 | Fringe patterns for shifted phase |

| Fig. 7-7 | Relationship between measured intensity and shifted phase |

spondence. In contrast, the methods shown in Fig. 7-2 and Fig. 7-4 are called passive stereo methods. The active stereo method using line patterns shown in Fig. 7-5
is widely applied in the industry. Various other measurement methods using structured light, such as two-dimensional lattice and random dot patterns, have been
proposed and implemented in practical settings.

　When a fringe pattern is used for structured light, the position on the pattern can
be determined with high accuracy using the phase-shift method. This method uses
fringes with a sinusoidal luminance pattern, as shown in Fig. 7-6. In this case, the
luminance I_0 at a certain point to be measured is expressed by Eq. (7-16), where I
is the amplitude of luminance change, θ is the phase, and I_{off} is the offset intensity.
The position on the pattern corresponds to phase θ. Therefore, the phase is obtained by intentionally changing (shifting) the phase of the projected fringes, taking multiple images, and combining the measurement results.

　Fig. 7-6 shows four fringe patterns with the phase shifted by $\pi/2$ rad. In this

の場合，位相を n 回シフトさせたときの輝度 I_n は式(7-17)で表される．このとき例え
ば，Fig. 7-6 に示すパターンの左端の地点での輝度の変化と位相シフト量の関係は
Fig. 7-7 の実線に示すとおりの正弦波状となり，測定された輝度 $I_0 \sim I_3$ はその正弦波
を位相シフト量間隔で離散的にサンプリングした点に対応する．求める θ は Fig. 7-7
に示す正弦波の初期位相と等しく，式(7-18) のように求められる．このように，4 つ
の位相での計測結果を組み合わせる方法を 4 位相法と呼び，式が簡単であるためによ
く用いられる．一般的には，離散フーリエ変換を用いて式(7-19) のように初期位相を
求めることができる．ここで，N は計測データの総数（位相シフト回数+1）である．

● 7.3　飛行時間測定法（TOF）

飛行時間測定法（TOF）では，Fig. 7-8 に示すように，カメラから計測対象物に光

$$I_0 = I\cos(\theta) + I_{\text{off}} \tag{7-16}$$

$$I_n = I\cos\left(\theta + \frac{n\pi}{2}\right) + I_{\text{off}} \tag{7-17}$$

$$\theta = \tan^{-1}\left(-\frac{I_3 - I_1}{I_2 - I_0}\right) \tag{7-18}$$

$$\theta = \tan^{-1}\left(-\frac{\displaystyle\sum_{n=0}^{N-1} I_n \sin\left(n\frac{2\pi}{N}\right)}{\displaystyle\sum_{n=0}^{N-1} I_n \cos\left(n\frac{2\pi}{N}\right)}\right) \tag{7-19}$$

case, the luminance I_n when the phase is shifted n times is expressed by Eq. (7-17).
For example, the relationship between the change in luminance at the leftmost
point of the pattern shown in Fig. 7-6 and the amount of phase shift is sinusoidal, as
shown by the solid line in Fig. 7-7. The measured luminance $I_0 - I_3$ correspond to
the points where the sinusoidal wave is discretely sampled at intervals of the given
phase shift. The required θ is equal to the initial phase of the sine wave shown in
Fig. 7-7 and can be obtained as in Eq. (7-18). This method by combining the measurement results at four phases is called the four-phase method and is often used
owing to its simple formulation. In general, the discrete Fourier transform can be
used to obtain the initial phase, as in Eq. (7-19), where N is the total number of
measurement data (number of phase shifts +1).

● 7.3　Time-of-flight（TOF）method

In the TOF method, as shown in Fig. 7-8, the light is irradiated from the camera
onto the object to be measured, and the distance to the object is calculated from the

を照射し，その光が反射して戻ってくるまでの時間から計測対象物までの距離を求める．計測対象物までの距離が求められると，7.1 節で述べた透視投影モデルを用いて XY 座標を求めることができる．TOF 法には大きく分けて 2 つの方法がある．1 つ目の方法は同図中に示すように光の照射と受光の間の遅れ時間 t_d を直接的に測定する直接 TOF 法である．この場合，計測対象物までの距離 d は式(7-20)を用いて求められる．ここで，c は光速である．直接法は遅れ時間 t_d を精度よく高分解能に測定する必要があり，計測器の実現は簡単ではない．

2 つ目の方法として間接 TOF 法がある．間接 TOF 法では，Fig. 7-9 に示すように，光の照射と同期して ON/OFF する受光素子（ここではゲートと呼ぶ）を用いること

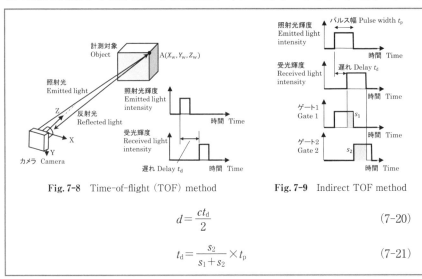

Fig. 7-8　Time-of-flight（TOF）method　　　　**Fig. 7-9**　Indirect TOF method

$$d = \frac{ct_d}{2} \tag{7-20}$$

$$t_d = \frac{s_2}{s_1 + s_2} \times t_p \tag{7-21}$$

time that it takes for the light to reflect back. TOF methods are divided mainly into direct and indirect methods. The direct TOF method directly measures the delay time t_d between the irradiation and reception of light, as shown in the figure. In this case, the distance d to the object to be measured is obtained using Eq. (7-20), where c is the speed of light. The direct method requires accurate and high-resolution measurement of the delay time t_d, making its implementation in a measuring instrument difficult.

In the indirect TOF method, as shown in Fig. 7-9, the time difference between light irradiation and light reception is determined using light-receiving elements (here called Gates) that turn on and off in synchronization with light irradiation. The time width of the ON/OFF of Gates 1 and 2 is set to be the same as the irradiation time t_d. Gate 1 is turned on at the same time as light irradiation. Gate 2 is

で，光の照射と受光の時間差を求める．ゲート 1, 2 が ON/OFF される時間幅は照射時間 t_p と同じに設定されている．ゲート 1 は光の照射と同じタイミングで ON される．ゲート 2 は光の照射が終了したタイミングで ON される．すると，2 つのゲートの受光量のバランスから，光の照射と受光の時間差を求めることができる．受光量に比例したゲート 1, 2 の信号強度をそれぞれ s_1, s_2 とすると，時間差 t_d は式(7-21)を用いて求められる．

間接 TOF 法は精度のよい時間計測が不要であり，比較的実用化が容易である．その反面，測定距離は制約される．Fig. 7-9 または式(7-21)からわかるように，$t_d \leqq t_p$ の時間差までしか計測できない．照射時間 t_p を長く設定することで測定距離を長くすることは可能であるが，環境光のノイズの影響を低減するためには t_p を短くして強度の高いパルス光を使いたい．ゲートの数を増やすことでも測定距離を長くすることができる．

● 7.4　縞歪み法

7.2 節で述べたステレオ法，7.3 節で述べた TOF 法では対象の 3 次元位置を計測する．また，両方法ともに対象表面は拡散反射が主となる光沢の少ない面である方が望ましく，鏡面の計測には不向きである．一方で，光学製品の計測などでは鏡面の計測

turned on at the time when light irradiation is completed. The time difference between light irradiation and light reception can then be determined from the balance of the amount of light received by the two gates. When the signal intensities of Gates 1 and 2, which are proportional to the amount of light received, are s_1 and s_2, respectively, the time difference t_d can be obtained using Eq. (7-21).

Although the indirect TOF method does not require precise time measurement and is relatively easy to implement, its measurement range is limited. As shown in Fig. 7-9 or Eq. (7-21), the time difference up to $t_d \leqq t_p$ can only be measured. Although the measurement range can be extended by increasing the irradiation time t_p, using a high intensity pulsed light with a short t_p is preferred in order to reduce the influence of ambient light noise. The measurement range can also be increased by increasing the number of gates.

● 7.4　Fringe deflectometry

The stereo method described in Section 7.2 and the TOF method described in Section 7.3 measure the 3D position of the object. Both methods are unsuitable for measuring specular surfaces because the surface of the object should be a low gloss

が必要とされる．そこで，鏡面を対象として，Fig. 7-10 に示すように対象の表面の傾きから形状を求める方法がある．計測点における表面の傾きを求める方法として Fig. 7-11 に示す縞歪み法がある．明暗で構成された縞パターンを計測対象に投影し，縞パターンの歪み方から表面の傾きを求める．Fig. 7-11 に示す例では，計測対象面が水平であり，正反射が起こるとすると，パターンの中央にある縞がカメラで撮影されるはずである．しかし，計測対象面が角度 α だけ傾くと，パターンの中央からずれた位置のパターンが撮影される．パターン位置のずれ量 d_θ と傾き α の関係は式(7-22)で表すことができるため，d_θ がわかれば α を求めることができる．輝度が正弦波状に変化する縞パターンを用いると，7.2 節で述べた位相シフト法によって高精度に d_θ を求めることができる．

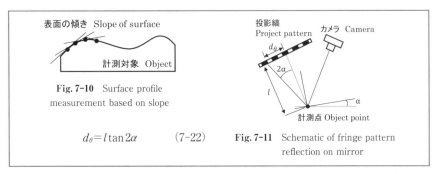

Fig. 7-10　Surface profile measurement based on slope

$$d_\theta = l \tan 2\alpha \qquad (7\text{-}22)$$

Fig. 7-11　Schematic of fringe pattern reflection on mirror

surface with diffuse reflection. In addition, the measurement of mirror surfaces is often required for the measurement of optical products. Therefore, a method exists for obtaining the shape of a mirror surface from the slope of the surface, as shown in Fig. 7-10. The fringe deflectometry shown in Fig. 7-11 is used to obtain the surface slope at the measurement point. In the example shown in Fig. 7-11, the fringe in the center of the pattern should be captured by the camera when the surface is horizontal and specular reflection occurs. However, when the measurement plane is tilted by an angle α, the camera captures the pattern at a position that is offset from the center of the pattern. The relationship between the offset of the pattern position d_θ and the tilt α can be expressed by Eq. (7-22), so α can be obtained if d_θ is known. When a fringe pattern with sinusoidal luminance variation is used, d_θ can be obtained with high accuracy using the phase-shift method described in Section 7.2.

【演習問題】

7-1) 平行ステレオにおける式(7-3)〜(7-5) が式(8-7)〜(8-9)と一致することを確かめよ.

7-2) 式(7-8) におけるカメラ行列は式(7-23)のように表される. ここで K はカメラの内部パラメータ行列, $[R|T]$ はワールド座標系からカメラ座標系への変換を表す行列である. Fig.7-2 に示す平行ステレオの場合, $[R|T]$ は式(7-24)〜(7-26) のように与えられる. このとき, 式(7-12)〜(7-15) の解が式(7-3)〜(7-5)と一致することを確かめよ.

7-3) Fig.7-2 に示す平行ステレオにおいて, 画像座標 $(x_L, y_L) = (7, 2)$ と $(x_R, y_R) = (-3, 2)$ が得られたとする. このとき, 式(7-12)〜(7-15)を最小二乗法を用いて解き, 点 A のワールド座標 (X_w, Y_w, Z_w) を求めよ. ただし $r=100, f=25$ とする.

7-4) TOF 法を用いた距離測定において, どのような誤差要因があるか述べよ.

$$P = K[R|T] \quad (7\text{-}23)$$

$$[R_L | T_L] = \begin{bmatrix} 1 & 0 & 0 & 0 \\ 0 & 1 & 0 & 0 \\ 0 & 0 & 1 & 0 \end{bmatrix} \quad (7\text{-}25)$$

$$K_L = K_R = \begin{bmatrix} f & 0 & 0 \\ 0 & f & 0 \\ 0 & 0 & 1 \end{bmatrix} \quad (7\text{-}24)$$

$$[R_R | T_R] = \begin{bmatrix} 1 & 0 & 0 & -r \\ 0 & 1 & 0 & 0 \\ 0 & 0 & 1 & 0 \end{bmatrix} \quad (7\text{-}26)$$

【Problems】

7-1) Verify that Eqs. (7-3)-(7-5) in parallel stereo agree with Eqs. (8-7)-(8-9).

7-2) The camera matrix in Eq. (7-8) is expressed by Eq. (7-23), where K is the intrinsic parameter matrix of the camera, $[R|T]$ is the transformation matrix from the world coordinates to the camera coordinates. In the case of the parallel stereo shown in Fig.7-2, K and $[R|T]$ are given as in Eqs. (7-24)-(7-26). In this case, verify that Eqs. (7-12)-(7-15) are consistent with Eqs. (7-3)-(7-5).

7-3) Suppose that image coordinates $(x_L, y_L) = (7, 2)$ and $(x_R, y_R) = (-3, 2)$ are obtained in the parallel stereo shown in Fig. 7-2. In this case, solve Eqs. (7-12)-(7-15) using the least-squares method to obtain the world coordinates (X_w, Y_w, Z_w) of point A where $r=100, f=25$.

7-4) Describe the error factors in measuring the distance to the object of measurement using the TOF method.

第8章　空間位置計測

　空間におけるある点の3次元位置を測定したい場合は多い．例えば，物体表面の点群それぞれの位置を測定することができれば，その物体の形状を求めて寸法誤差や形状誤差の評価が可能となる．また，ロボットハンドの先端など，運動している点の位置を測定することは，運動誤差の評価や補正に寄与する．空間位置は一般的に直交座標系における XYZ 座標を用いて表される．空間位置の測定方法には，測定対象の XYZ 座標を直接的に測定する方法と，距離と偏角を測定して極座標を求めて直交座標に変換する方法などの間接的な測定方法がある．ここでは，代表的な4つの測定方法について説明する．

● 8.1　直交座標測定系
　空間位置を測定する最も単純な方法として，直交座標系における XYZ 座標を直接

Chapter 8　Measurement of Volumetric Position

　Measuring the three-dimensional position of a point in space is often required. For example, if the positions of a group of points on the surface of an object can be measured, the geometry of the object can be determined to evaluate geometric errors. Measuring the position of a point in motion, such as the tip of a robot hand, can also contribute to the evaluation and correction of motion errors. Spatial position is generally expressed using XYZ coordinates in a Cartesian coordinate system. Two methods for spatial position measurements are direct measurement of the XYZ coordinates of the object to be measured and indirect measurement methods such as measuring the distance and declination in the polar coordinate system and converting them to the Cartesian coordinate system. Four representative measurement methods are described here.

● 8.1　Cartesian measurement system
The simplest method of measuring spatial position is to directly measure XYZ

的に測定する方法がある．Fig. 8-1 に直交座標測定の模式図を示す．原点 O からの座標 x, y, z を測定することで，測定対象点 P の座標 (p_x, p_y, p_z) は式(8-1)～(8-3)のように求められる．

　本方法を用いた代表的な計測装置として Fig. 8-2 に示す三次元測定機がある．三次元測定機はプローブを移動させるための直交 3 軸の送りと位置検出器を備えている．プローブを測定対象物の表面に接触させて，プローブの位置を位置検出器で測定することで，接触点の座標を求めることができる．直交座標測定系での測定は原理が単純であり，わかりやすい．しかし，測定対象点と位置検出器が離れてしまうことが多い

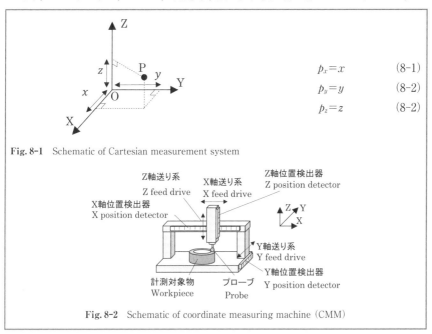

$$p_x = x \qquad (8\text{-}1)$$

$$p_y = y \qquad (8\text{-}2)$$

$$p_z = z \qquad (8\text{-}2)$$

Fig. 8-1　Schematic of Cartesian measurement system

Fig. 8-2　Schematic of coordinate measuring machine（CMM）

coordinates in the Cartesian coordinate system. Fig. 8-1 shows a schematic diagram of the Cartesian coordinate measurement system. By measuring the coordinates x, y, z, from origin O, the coordinates of point P(p_x, p_y, p_z) can be obtained as Eqs. (8-1)-(8-3). A typical measurement device using this method is a coordinate measuring machine（CMM）shown in Fig. 8-2. The CMM is equipped with three-axis orthogonal feed drives to move the probe and position detectors. The probe contacts the surface of the object to be measured, and the position of the probe is measured by the position detector to obtain the coordinates of the point of contact. The principle of measurement in the Cartesian coordinate measuring system is simple and easy to understand. However, measurement errors (e.g., Abbe errors, see

ため，測定系の変形による測定誤差（例：アッベ誤差，11章参照）が生じやすい．

● 8.2　極座標測定系

距離と偏角を測定して測定対象物の極座標を求め，直交座標に変換することで空間
位置を計測できる．Fig.8-3 に極座標測定の模式図を示す．原点 O からの距離 l と，偏
角 ϕ, θ を測定することで，測定対象点 P の座標 (p_x, p_y, p_z) は式(8-4)〜(8-6) のよ
うに求められる．

本方法を用いた代表的な測定器として Fig.8-4 に示すようなレーザを用いた距離計
（例：レーザ干渉計）と 2 軸の回転機構および角度検出器を組み合わせた測定器があ
り，レーザトラッカ，レーザレンジファインダなどの名称で商品化されている．本測
定器はレーザを測定対象物に向けることで，距離計を用いて測定対象物までの距離を

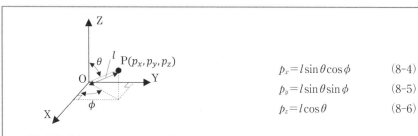

$$p_x = l \sin\theta \cos\phi \qquad (8\text{-}4)$$

$$p_y = l \sin\theta \sin\phi \qquad (8\text{-}5)$$

$$p_z = l \cos\theta \qquad (8\text{-}6)$$

Fig. 8-3　Schematic of polar coordinate
measurement system

Chapter 11) are likely to occur due to deformation of the measurement system because the point to be measured and the position detector are often far apart.

● 8.2　Polar coordinate measurement system

In the polar coordinate measurement system, the distance and declination are
measured to obtain the polar coordinates of the object to be measured. Then, the
measured polar coordinates are converted to Cartesian coordinates to obtain their
spatial position. By measuring the distance l from the origin O and the declination
angles ϕ and θ, the coordinates (p_x, p_y, p_z) of point P can be obtained by Eqs. (8-4)-
(8-6).

A typical measuring instrument using this method combines a distance meter
using a laser (e.g., laser interferometer) with 2-axis rotational drives and angle detectors, as shown in Fig. 8-4. Such a device is commercialized as a laser tracker and
a laser rangefinder. The laser rangefinder measures the distance to the object to be
measured using a laser interferometer and measures the declination angle using

測定し，角度検出器を用いて偏角を測定する．2軸の回転機構を用いて走査すること
で，衝突回避のための物体検出や測定対象物の形状測定が行える．小型の測定器であっ
ても広い測定範囲が得られるため，様々な分野で応用されている．式(8-4)～(8-6)か
らわかるように，点 P の座標 (p_x, p_y, p_z) に距離 l が含まれるため，l が大きくなると，
測定分解能と測定精度に角度測定の分解能と精度が与える影響が大きくなる．

● 8.3　三角測量系

既知の距離離れた2点から計測対象点への方向がわかっている場合，三角測量の原
理を用いて空間位置を求めることができる．Fig. 8-5 に三角測量を用いた空間位置測
定の模式図を示す．点 P が測定対象点であるとし，原点 O と点 A に測定器が設置さ
れているとする．線分 OA の長さ r を基線長と呼び，予め測定しておく．また，点 P

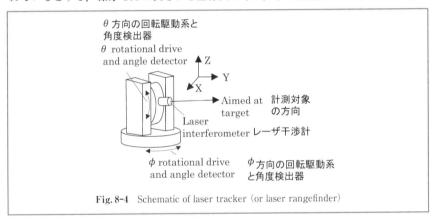

Fig. 8-4　Schematic of laser tracker (or laser rangefinder)

angle detectors by directing a laser beam toward the object to be measured. The
system is widely applied because a wide measurement range can be obtained even
with a small measuring instrument. As shown in Eqs. (8-4)-(8-6), because the coordinates (p_x, p_y, p_z) of point P include the distance l, when l is large, the influence of
the resolution and accuracy of angle measurement on the measurement resolution
and accuracy increases.

● 8.3　Triangulation system

Using the principle of triangulation, a spatial position can be obtained by measuring the orientations from two points at a known distance from each other. Fig. 8-5
shows a schematic diagram of spatial position measurement using triangulation.
Assume that point P is the point to be measured, and that measuring instruments
are placed at the origin O and point A. The length of the line segment OA, r, is

をXY平面に投影した点をP′とする．線分OP′と線分AP′がX軸とそれぞれ成す角度, ϕ_1, ϕ_2 と, 線分OPと線分APがXY平面とそれぞれなす角度 θ_1 もしくは角度 θ_2 を測定することで, 点Pの座標 (p_x, p_y, p_z) は式(8-7)〜(8-10)のように求められる．

式(8-7)〜(8-10)の導出のために, Fig. 8-6に示す模式図を用いて, 点P′の2次元座標を求める場合を説明する．基線長 r は式(8-11)で表すことができる．式(8-11)に三角関数の公式である式(8-12), (8-13)を代入することで, 式(8-8)と同様に p_y を求めることができる．さらに, 式(8-14)から, 式(8-7)と同様に p_x が求められる．すると, 線分OP′と線分AP′の長さから, 式(8-9)と(8-10)をそれぞれ導出できる．

三角測量では測定対象までの距離を測定する必要がないため, 遠方の測定対象であっても比較的簡単に位置を求めることができる．このため, 古くから土地の測量や構造物の高さを求める際などに利用されてきた．現在においても, 第7章で述べてい

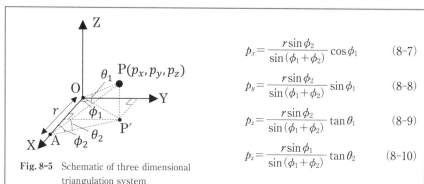

$$p_x = \frac{r \sin\phi_2}{\sin(\phi_1+\phi_2)} \cos\phi_1 \tag{8-7}$$

$$p_y = \frac{r \sin\phi_2}{\sin(\phi_1+\phi_2)} \sin\phi_1 \tag{8-8}$$

$$p_z = \frac{r \sin\phi_2}{\sin(\phi_1+\phi_2)} \tan\theta_1 \tag{8-9}$$

$$p_z = \frac{r \sin\phi_1}{\sin(\phi_1+\phi_2)} \tan\theta_2 \tag{8-10}$$

Fig. 8-5　Schematic of three dimensional triangulation system

called the baseline length and is measured in advance. Let P′ be the projection of point P onto the XY plane. The coordinates of point P (p_x, p_y, p_z) can be obtained by Eqs. (8-7)-(8-10), where ϕ_1 is the angle between the lines OP′ and OA, ϕ_2 is the angle between the lines AP′ and OA, θ_1 is the angle between the lines OP′ and OP, and θ_2 is the angle between the lines AP′ and AP.

To derive Eqs. (8-7)-(8-10), we will use the schematic diagram shown in Fig. 8-6 to explain the case of obtaining the two-dimensional coordinates of point P′. The baseline length r can be expressed by Eq. (8-11). Substituting the trigonometric formulas (8-12) and (8-13) into Equation (8-11), p_y can be obtained similarly to Eq. (8-8). Furthermore, from Eq. (8-14), p_x can be obtained similarly to Eq. (8-7). Then, Eqs. (8-9) and (8-10) can be derived from the lengths of line OP′ and line AP′, respectively.

Triangulation does not require measuring the distance from the object being measured, making it relatively easy to determine the position of a distant measure-

るステレオ法は三角測量の原理に基づいており，応用範囲の広い測定法といえる．角度の測定誤差がある場合を考えると，3次元での測定では Fig. 8-5 における点 O，A と測定対象を結ぶ直線が交わらない，具体的には，理想的には一致するはずの式(8-9)，(8-10)が一致しなくなる．このため，点 O，点 A からの2つの直線の垂線の中間点を求める方法などが用いられる．

● 8.4 多辺測量系

多辺測量の原理を用いて，座標のわかっている3点以上の地点から測定対象までの距離を測定することで測定対象の空間位置を求めることができる．Fig. 8-7 に3点からの距離を用いた三辺測量の模式図を示す．点 P が測定対象点であるとする．距離測定の基点となる3点（つまりセンサの位置）は3次元空間上で任意に配置できるが，Fig. 8-7 に示すように原点 O，X 軸上の点 A，XY 平面上の点 B に基点の3点をそれ

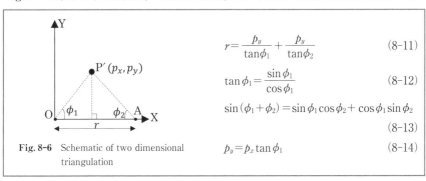

$$r = \frac{p_y}{\tan\phi_1} + \frac{p_y}{\tan\phi_2} \qquad (8\text{-}11)$$

$$\tan\phi_1 = \frac{\sin\phi_1}{\cos\phi_1} \qquad (8\text{-}12)$$

$$\sin(\phi_1 + \phi_2) = \sin\phi_1\cos\phi_2 + \cos\phi_1\sin\phi_2 \qquad (8\text{-}13)$$

$$p_y = p_x\tan\phi_1 \qquad (8\text{-}14)$$

Fig. 8-6 Schematic of two dimensional triangulation

ment object. For this reason, it has long been used for surveying land and determining the height of structures. The stereo method described in Chapter 7 also operate on the basic triangulation principle. Thus, triangulation remains a measurement method with varying applications. Considering the measurement error, the lines from points O and A often do not intersect in 3D measurements, and the least-squares method is used to obtain the most reliable position.

● 8.4 Multilateration system

Multilateration determines the spatial position of an object only by distance from base points to the object. Fig. 8-7 shows a schematic diagram of multilateration with three base points. Let point P be the point to be measured. The three base points for distance measurement (i.e., sensor positions) can be placed arbitrarily in a three-dimensional space. However, as shown in Fig. 8-7, the three points can be placed at origin O, point A on the X axis, and point B on the XY plane, respectively,

ぞれ配置しても一般性を失わない．各基点から点 P までの距離を l_1, l_2, l_3 とすると，点 P は各基点を中心とする球面上にある．つまり，式(8-15)〜(8-17)が成り立つ．ここで，点 A，B の座標 x_2, x_3, y_3 が既知であり，測定によって l_1, l_2, l_3 が得られたとすると，式(8-15)〜(8-17)の連立方程式を解くことで点 P の座標 (p_x, p_y, p_z) が求められる．

多辺測量の身近な例に GPS や Wi-Fi を用いたスマートフォンの位置推定がある．例えば Fig. 8-8 に示すように，3 点の Wi-Fi アクセスポイントからの電波強度から距離を求め，多辺測量の原理で位置を推定する．しかし，実際の測定では距離 l_1, l_2, l_3 には測定誤差が含まれる．すると，未知数が点 P の座標 3 つだけではなくなるため，より多数の基点からの距離測定が必要となる．例えば，各距離に同じ誤差が含まれる場合には未知数は 4 つとなり，4 つの基点からの距離測定が必要である．さらに，マイクロメートルレベルの計測精度が必要な高精度測定の場合では，基点の位置が十分な

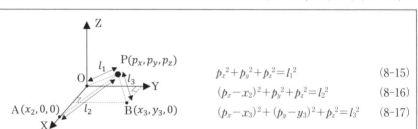

$$p_x{}^2 + p_y{}^2 + p_z{}^2 = l_1{}^2 \qquad (8\text{-}15)$$

$$(p_x - x_2)^2 + p_y{}^2 + p_z{}^2 = l_2{}^2 \qquad (8\text{-}16)$$

$$(p_x - x_3)^2 + (p_y - y_3)^2 + p_z{}^2 = l_3{}^2 \qquad (8\text{-}17)$$

Fig. 8-7　Schematic of multilateration with three base points

without loss of generality. When the distances from each base point to point P are l_1, l_2, l_3, then point P is on a sphere centered at each base point. In other words, Eqs. (8-15)-(8-17) hold. Assuming that the coordinates x_2, x_3, and y_3 of points A and B are known, and that the distances l_1, l_2, and l_3 are obtained by measurement, the coordinates (p_x, p_y, p_z) of point P can be obtained by solving the simultaneous equations of Eqs. (8-15)-(8-17).

A familiar example of multilateration is the position estimation of a smartphone using GPS and Wi-Fi, as shown in Fig. 8-8. The distance is determined from the radio wave intensity from three Wi-Fi access points, and the position is estimated using the principle of multilateration. However, in actual measurements, the distances l_1, l_2, and l_3 contain measurement errors. Then, the number of unknown parameters exceeds the three coordinates of point P. Therefore, increasing the number of base points is necessary. For example, when each distance contains the same error, the number of unknown parameters is four. Thus, distance measurements from four

精度ではわからない場合も多い．この場合は計測結果から点 P の座標と各基点の座標を同時に同定する必要がある．Fig. 8-9 に 4 つの基点からの距離測定による多辺測量の模式図を示す．基点の数を 4 つに増やしたとしても，未知パラメータの数は基点の座標 6 つと計測対象点の座標 3 つの合計 9 つとなり，方程式の数が足りない．そこで，計測点の数も増やして m 点で計測を行うとすると，未知パラメータの数＝$6+3m$，方程式の数＝$4m$ となり，$m \geq 6$ の場合に「方程式の数 ≧ 未知パラメータの数」となる．このような計測において，非線形最小二乗法を用いた反復計算によって計測点と基点の位置を同時に同定する方法が提案されている．

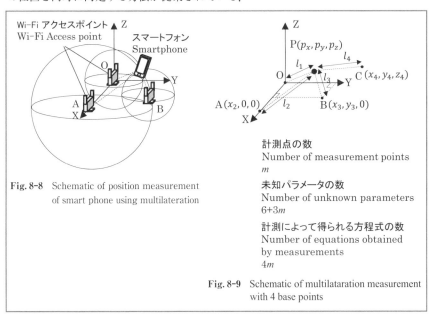

Fig. 8-8 Schematic of position measurement of smart phone using multilateration

計測点の数
Number of measurement points
m

未知パラメータの数
Number of unknown parameters
$6+3m$

計測によって得られる方程式の数
Number of equations obtained by measurements
$4m$

Fig. 8-9 Schematic of multilataration measurement with 4 base points

base points are required. Furthermore, in the case of high-precision measurements requiring micrometer-level measurement accuracy, the position of the base point is often unknown with sufficient accuracy. Fig. 8-9 shows a schematic diagram of a multilateration with four base points. Even if the number of base points is increased to four, the number of unknown parameters is nine, comprising six coordinates of base points and three coordinates of points to be measured. If the number of measurement points is also increased to m, the number of unknown parameters=$6+3m$ and the number of equations=$4m$. So, when $m \geq 6$, the number of unknown parameters is less than the number of equations. In such cases, a method has been used to simultaneously identify the positions of the measurement point and the base point by iterative computation using a nonlinear least-squares method.

【演習問題】

8-1) Fig. 8-2 に示す三次元測定機において，アッベ誤差（第 11 章参照）を生じない
ようにする方法を述べよ．

8-2) Fig. 8-3 に示す極座標測定系での角度測定において，標準不確かさ $u(\phi)=1\times$
10^{-4} rad の場合に，p_x の合成標準不確かさを求めよ．ただし，$\theta=\pi/2$ rad, $l=$
1 m とする．

8-3) Fig. 8-5 に示す三角測量系での角度測定において，標準不確かさ $u(\phi_1)=u(\phi_2)$
$=1\times10^{-4}$ rad の場合に，p_x の合成標準不確かさを求めよ．ただし，$\theta_1=\theta_2=$
0 rad, $r=0.5$ m, $\pi/6$ rad$\leqq\phi_1\leqq\pi/3$ rad, $\pi/6$ rad$\leqq\phi_2\leqq\pi/3$ rad, とする．合成
標準不確かさは ϕ_1 と ϕ_2 に依存するため，表計算ソフトウエアなどを用いて ϕ_1
と ϕ_2 と合成標準不確かさの関係を表す表を作成するのがよい．

8-4) Fig. 8-7 に示す 3 つの基点を用いた多辺測量系において，$x_2=x_3=y_3=l_1=l_2=l_3$
$=100$ mm とするとき，式(8-15)〜(8-17)を用いて点 P の座標 (p_x, p_y, p_z) を求
めよ．

【Problems】

8-1) Describe how to avoid the Abbe error (see Chapter 11) on the CMM shown
in Fig. 8-2.

8-2) Find the combined standard uncertainty of p_x in the polar coordinate measuring system shown in Fig. 8-3 when the standard uncertainty for angular measurement $u(\phi)=1\times10^{-4}$ rad. However, let $\theta=\pi/2$ rad and $l=1$ m.

8-3) Find the composite standard uncertainty of p_x in the triangulation system
shown in Fig. 8-5 when the standard uncertainty for an angle measurement u
$(\phi_1)=u(\phi_2)=1\times10^{-4}$ rad. Assume that $\theta_1=\theta_2=0$ rad and $r=0.5$ m, $\pi/6$ rad$\leqq$
$\phi_1\leqq\pi/3$ rad, $\pi/6$ rad$\leqq\phi_3\leqq\pi/3$ rad. Because the combined standard uncertainty depends on ϕ_1 and ϕ_2, create a table showing the relationship between ϕ_1, ϕ_2
and the combined standard uncertainty using spreadsheet software, etc.

8-4) In multilateration with the three base points shown in Fig. 8-7, when $x_2=x_3=$
$y_3=l_1=l_2=l_3=100$ mm, find the coordinates (p_x, p_y, p_z) of point P using Eqs.
(8-15)-(8-17).

第9章 光学顕微鏡

光学顕微鏡は，光学拡大により測定対象を拡大観察するために用いられる．デジタル画像処理技術の発展により光学観察技術への用途は拡大し，精密計測に留まらず，基礎科学から産業応用まで幅広い領域で欠かせない技術となっている．画像処理技術は重要であるが，そもそも，高品質画像を取得する光学顕微法の重要性はいうまでもない．加えて，光学顕微鏡は，例えば色収差や焦点深度など，光学特性を駆使することで測定物の表面位置センシングなどが可能で，拡大観察を超えた微細構造や表面性状の計測技術へと広く発展している．そこで本章では，光学的な顕微システムの基本構成や画像取得を理解するための基本的な用語についてまとめ，また，光学顕微法を応用した表面形状計測技術について説明する．

Chapter 9　Optical Microscopy

Optical microscopes magnify and observe the measured object based on optical magnification. Digital image processing technologies, such as pattern matching and machine learning, have expanded their applications. Nowadays, it has become indispensable not only for precision measurement but also in various fields, from science to industrial applications. It is important for optical microscopes to obtain high quality images, as this facilitates the sensitive detection of a measured object's surface position. This is achieved by leveraging diverse optical characteristics such as chromatic aberration and depth of focus. Many types of measurement techniques for microstructures and surface textures integrated into optical microscopes have been developed in the field of precision measurement. Therefore, this chapter summarizes the fundamental terms used to understand the basics of optical microscopy systems and image acquisition and also describes optical microscopy-based surface measurement techniques.

● 9.1　光学顕微鏡の基礎的構成要素

光学顕微鏡の基礎的な構成要素を Fig. 9-1 に示す．測定対象を照らす光源および照明系，測定対象を光学拡大する光学観察系，その光学応答を取得するイメージセンサなどの検出器，およびその取得信号を処理する計算機が主な構成要素である．

● 9.2　光学顕微鏡に関する基礎用語

光学顕微鏡を理解する上で基礎となる用語についてまずは整理する．

結像倍率　顕微鏡の主目的は物体の拡大観察であり，レンズ系を用いて拡大する．レンズによる光学拡大を Fig. 9-2 に示すが，レンズから距離 s_o にある物体が，距離 s_i で

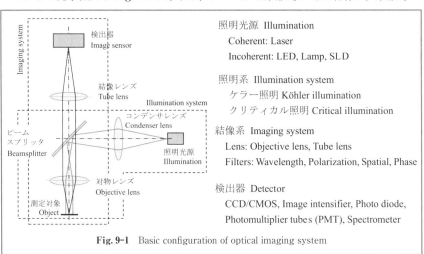

照明光源　Illumination
　　Coherent: Laser
　　Incoherent: LED, Lamp, SLD

照明系　Illumination system
　　ケラー照明 Köhler illumination
　　クリティカル照明 Critical illumination

結像系　Imaging system
　　Lens: Objective lens, Tube lens
　　Filters: Wavelength, Polarization, Spatial, Phase

検出器　Detector
　　CCD/CMOS, Image intensifier, Photo diode,
　　Photomultiplier tubes (PMT), Spectrometer

Fig. 9-1　Basic configuration of optical imaging system

● 9.1　Basic structures of optical microscope

The basic components of an optical microscope are shown in Fig. 9-1. The main components are a light source, an illumination system, an optical observation system that optically magnifies the object, a detector that acquires its optical response, and a computer that processes the acquired signals.

● 9.2　Fundamental terms related to optical microscopy

Some fundamental terms for understanding optical microscopy are summarized here.

Magnification　The purpose of an optical microscope is to magnify an object using a lens system. Fig. 9-2 shows optical magnification using a lens. An object at a dis-

結像している．焦点距離を f とすると，これらの位置関係はレンズの公式を用いて式 (9-1) のように表される．形成された像サイズ H' と物体サイズ H の比が光学系の横倍率 M となる（式(9-2)）．レンズを組み合わせて倍率を上げることで像サイズを拡大し，物体を顕微観察する．

開口数（NA） 光学システムの伝達特性を決定づける上で重要となるのがレンズの開口数（NA）である．NA は式(9-3)で表される．Fig.9-3 で示すように，レンズで集光する際，最も外側の光線が光軸となす角の大きさが θ であり，n は媒質の屈折率である．NA が大きいほど，光は物体に対して大きな傾斜角で入射でき，同時に，物体からの光を広い範囲で集光できる．そのような幾何学的な構成から高 NA の対物レンズ

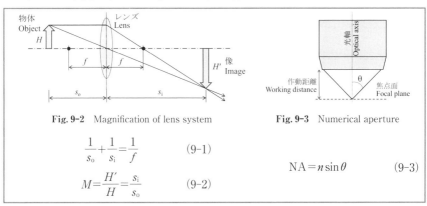

Fig. 9-2 Magnification of lens system

Fig. 9-3 Numerical aperture

$$\frac{1}{s_o}+\frac{1}{s_i}=\frac{1}{f} \qquad (9\text{-}1)$$

$$M=\frac{H'}{H}=\frac{s_i}{s_o} \qquad (9\text{-}2)$$

$$NA=n\sin\theta \qquad (9\text{-}3)$$

tance s_o from the lens forms an image at a distance s_i. The relationship between s_o and s_i is expressed as the focal length of lens f, using the lens formula as in Eq.(9-1). The ratio of the formed image size H' to the object size H is the lateral magnification factor M of the optical system, as shown in Eq.(9-2). By compounding lenses to increase magnification, the image size is enlarged, and the microscopic image of the object is observed.

Numerical aperture The numerical aperture (NA) of a lens plays an important role in optical microscopy imaging, as it determines optical transmission characteristics. NA is defined by Eq.(9-3). As shown in Fig.9-3, θ is the angle of the outermost optical ray and the optical axis for the focused beam, and n is the refractive index of the medium. A larger NA results in a larger incident angle. Concurrently, the objective lens can collect light over a wider angular range. A higher NA of an objective lens results in a smaller distance between the focal plane and objective front lens. This distance between the focal plane and the objective front lens is

ほど，レンズ端面から焦点面が近くなる．この距離を作動距離と呼ぶ．

回折限界　光学顕微鏡の結像性能は点像分布で考えることができる．ある点光源を理想光学系で観察すると，結像面ではある幅を持った像（点像分布）が観測される（Fig. 9-4）．この点像の電場振幅分布は式（9-4）で与えられる．ただし，$J_1(v)$ は 1 次の第 1 種ベッセル関数である．光学システムの集光性能は，例えば，観察された点像分布の半値全幅（FWHM）d と考えることができ，レンズ NA と光波長 λ によって式（9-5）のように表される．FWHM の他にも $1/e^2$ などで評価される場合もある．このような光の回折現象に伴う集光限界による，物体の解像限界を回折限界という．回折限界は多くの場合「その点像分布の極小点と他点によって形成される点像分布の極大点を重ねたときが解像可能な最小距離（Fig. 9-4）」というレイリー基準の考え方に基づいて与

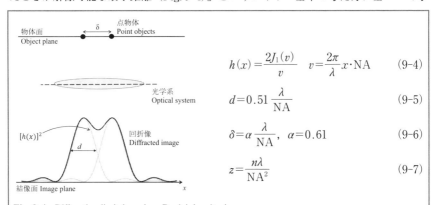

$$h(x) = \frac{2J_1(v)}{v} \quad v = \frac{2\pi}{\lambda}x \cdot \mathrm{NA} \qquad (9\text{-}4)$$

$$d = 0.51\frac{\lambda}{\mathrm{NA}} \qquad (9\text{-}5)$$

$$\delta = \alpha\frac{\lambda}{\mathrm{NA}}, \quad \alpha = 0.61 \qquad (9\text{-}6)$$

$$z = \frac{n\lambda}{\mathrm{NA}^2} \qquad (9\text{-}7)$$

Fig. 9-4　Diffraction limit based on Rayleigh criterion

called the working distance.

Diffraction limit　The imaging performance of an optical microscope can be considered in terms of the point spread function (PSF). As shown in Fig. 9-4, when a point source is observed, an image with a certain width (i.e., PSF) is observed on the image plane. The electric field amplitude distribution of the PSF is given by Eq. (9-4), where $J_1(v)$ is a first-order Bessel function of the first kind. If considering the FWHM of the PSF, d, as the width of the focal spot, the focusing performance of an optical system is determined by NA and wavelength λ, which is expressed as in Eq. (9-5). In addition to FWHM, it may also be evaluated as $1/e^2$, etc. The focusing limit caused by the optical diffraction phenomenon is called the diffraction limit. The diffraction limit is often given based on the Rayleigh criterion that "the minimum distance at which an image can be resolved is the distance between the first mini-

えられ，式(9-6)で表される．

ここで，光学系を情報伝達システムと考える．光学システムによって伝達できる空間周波数には制限があり，回折限界を超えるほどに密に存在する物体情報の高い空間周波数は伝達できない，ともいえる．しかし，通常の一様照明ではなく，照明光強度に空間分布を持たせる（例えば縞照明）ことでモアレ効果による空間周波数シフトを起こし，同じ光学システムでも，より高い周波数成分の信号を取得できる．つまり，より近くに隣接した点物体を見分けることができる．他にも非線形光学現象を用いて回折限界を上回るレベルの解像度を得る超解像技術が開発されている．

焦点深度　光学顕微鏡において，物体の表面が対物レンズの焦点面から外れると，得られる像にボケが生じる．このボケがほぼ無視できる焦点面の光軸方向範囲を焦点深度といい，式(9-7)で表される．焦点深度が深いと，光軸方向に変化の大きい物体でも全体的に焦点のあった像が得られる．

mum point of the distribution and the maximum point of the diffracted image formed by the other points (Fig. 9-4)," and is expressed by Eq.(9-6).

Consider the optical system as an information transfer system. An optical system can transmit a limited spatial frequency, and the high spatial frequency of object information that is dense enough to exceed the diffraction limit cannot be transmitted. However, instead of the usual uniform illumination, a spatial distribution of illumination intensity (e.g., structured illumination) causes a spatial frequency shift due to the moiré effect, and the same optical system can acquire signals with higher frequency components. In other words, point objects that are more closely adjacent can be distinguished. Additionally, alternative super-resolution methods that surpass the diffraction limit have been developed, utilizing nonlinear optical phenomena to achieve enhanced resolution.

Depth of focus　In optical microscopy, displacing an object's surface from the focal plane of the lens system results in a blurred image. The range in the optical axis direction of the focal plane where the image blur is almost negligible is called the depth of focus, which is expressed by Eq.(9-7). When the depth of focus is wide, an entirely focused image can be obtained even for an object with a large change in the optical axis direction.

収差 結像系を光線で考える．1点から出た光を結像する際に，像面で1点に結ばない現象を収差と呼ぶ．光軸近傍かつ光軸との交差角が小さい光線を近軸光線と呼び，結像を解析する光線追跡計算では，近軸近似が用いられる．しかし現実の光学系では，近軸での取り扱いでは不都合が生じ，収差が現れる．収差には，大きく分けてザイデルの5収差と，色収差がある．色収差は，レンズ材料の屈折率が波長によって異なることに起因する収差である（Fig. 9-5(a)）．一方，単一波長でも発生する単色収差がザイデルの5収差と称される（Table 9-1，Fig. 9-5(b)〜(f)）．

アッベの正弦条件 光学顕微鏡やそれを用いた計測では，歪みのない像取得が重要である．物体が物体面内のどこにあっても等しい点像分布で，像面に結像する必要がある．そのための条件が正弦条件である．Fig. 9-6 に示すように，物体面に小さい開口が周期 Λ で並んでいる場合を考える．そこに光軸に平行な平面波を入射すると，0次光（透過光）と m 次回折光が生じる．回折光は角度 θ_m 方向に進み（式(9-8)），光学系によって集光され角度 θ'_m で像面に照射される．像面で形成される開口像の周期 Λ'_m は，式(9-9) となる．各回折次数 m に対して，結像倍率 $\Lambda'_m/\Lambda(=M)$ は等しくなけ

Aberration Considering ray optics in an imaging system, the phenomenon in which light emitted from a single point does not form a single point on the image plane is called aberration. Rays that are close to the optical axis and have a small angle with the optical axis are called paraxial rays, and paraxial approximation is used in ray-tracing calculations to analyze the imaging system. However, the practical phenomenon deviates from the paraxial approximation, and aberrations appear. Two main types of aberrations are five Seidel and chromatic aberrations. Chromatic aberration occurs because the refractive index of the lens material varies with the wavelength (Fig. 9-5(a)). Five Seidel aberrations can occur even at a single wavelength, a monochromatic aberration. Explanations are provided in Table 9-1, and illustrations are shown in Fig. 9-5(b) to (f).

Abbe's sine condition To ensure precise measurement, obtaining a distortion-free image is important. Any point in the object plane must be imaged on the image plane with the same PSF. As shown in Fig. 9-6, consider the case where small apertures are aligned on the object plane with period Λ. When a plane wave is an incident parallel to the optical axis, 0th-order light (transmitted light) and mth-order diffracted light occur. The diffracted light travels in the direction of angle θ_m (Eq. (9-8)), which is focused by the optical system, and is then irradiated onto the image

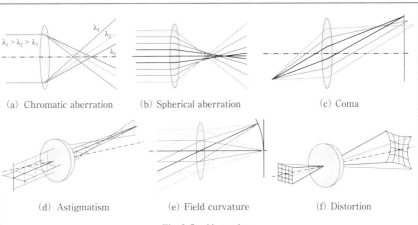

(a) Chromatic aberration   (b) Spherical aberration   (c) Coma

(d) Astigmatism   (e) Field curvature   (f) Distortion

Fig. 9-5   Aberrations

Table 9-1   Five Seidel aberrations

球面収差 Spherical aberration	非近軸光線の焦点ずれによる収差．これを解消するため，非球面レンズが用いられる． Aberration caused by out of focus due to non-paraxial rays. Aspherical lenses are used to eliminate.	Fig. 9-5(b)
コマ収差 Coma	レンズ主平面が，現実的には曲面であることから，光軸外の物体を結像した際，円錐状にボケが生じる． Coma causes a conical blur when an object is at off-axis, which is because the principal plane of the lens is a curved.	Fig. 9-5(c)
非点収差 Astigmatism	光軸外の物体を結像する際，レンズの円周方向と半径方向で，集光位置が異なることによる収差． Aberration caused by the difference in focuses between the circumferential and radial directions of the lens when an object is at off-axis.	Fig. 9-5(d)
像面収差 Field curvature	理想的に平面として扱う像面や物体面が，現実には曲面であることによる収差． Field curvature is an aberration caused by the fact that the image or object plane is ideally treated as a plane but is curved practically.	Fig. 9-5(e)
歪曲収差 Distortion	近軸領域から外れると，結像倍率がレンズの各部によって異なることで，像が歪む． In the non-paraxial region, the image is distorted because the magnification varies in the plane.	Fig. 9-5(f)

ればならない．等しくない場合，回折次数によって結像位置が異なるという収差が発生する．よって，結像倍率 Λ'_m/Λ が等しくなる式(9-10)をアッベの正弦条件という．

● 9.3　合焦点顕微法

光学顕微鏡は物体面近傍の情報を結像面に光学伝達するシステムでもあり，観察だけでなく計測や分析などにも広く展開されている．精密計測においては，測定対象の拡大観察や，設計図面の縮小投影などに用いられる．それに留まらず，物体面での光学応答取得を利用した表面形状の計測などにも応用されている．ここではいくつかの

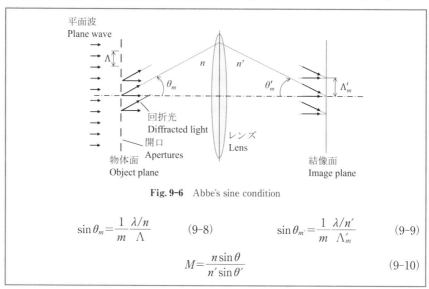

Fig. 9-6　Abbe's sine condition

$$\sin\theta_m = \frac{1}{m}\frac{\lambda/n}{\Lambda} \qquad (9\text{-}8)$$

$$\sin\theta_{m'} = \frac{1}{m}\frac{\lambda/n'}{\Lambda'_m} \qquad (9\text{-}9)$$

$$M = \frac{n\sin\theta}{n'\sin\theta'} \qquad (9\text{-}10)$$

plane at angle θ'_m. The period Λ'_m of the aperture image formed at the image plane is given by Eq.(9-9). For each diffraction order m, the magnification Λ'_m/Λ must be equal. If they are unequal, aberrations occur in which the image positions differ depending on the diffraction order. Eq.(9-10) is the so-called Abbe's sine condition.

● 9.3　Focus variation microscopy

An optical microscope is a system that optically transmits information near the object plane to the imaging plane. Therefore, the optical microscope system is widely used for observation, measurement, and analysis at local points. In precision measurement, it is used, for example, to observe measured objects and to project designed semiconductor circuits. It is also widely used for surface profile measurement. Here, several representative dimensional measurement methods are

代表的な手法を紹介する．まず，合焦点法を用いた手法について述べる．

　光学顕微鏡系では，物体面と像面は共役関係にある．合焦点法では，最も焦点の合う位置（合焦点位置）をイメージセンサの画像情報をもとに探索することで，物体表面の位置を検出する．物体を光軸方向に走査し，面内すべての点で合焦点位置を探索すると，物体の高さ分布を計測できる．合焦点の判定は，微分処理などで得られる画像コントラスト分布を用いる．コントラストが最も高い光軸方向の位置を合焦点位置と判定し，これを測定物の表面とする．この手法は合焦点法，Focus variation 法，Shape from focus 法などと呼ばれる．横分解能は，回折限界によって規定され，光軸方向分解能は焦点深度に依存する．後述の共焦点法や白色干渉法よりも光軸方向分解能は劣るが，測定系がシンプルであり，幅広い測定対象を計測可能である．注意すべきは，画像コントラストが得にくい滑らかな表面や鏡面の計測は難しい．

● 9.4　共焦点顕微法

走査型共焦点顕微鏡は，マイクロスケールの微細形状を計測する汎用的測定器であ

introduced. The focus variation-based method is first described.

The object and image surfaces are conjugated in an optical microscope system. In the focus variation method, the position of the object surface is detected by searching for the most in-focus position (focal point) based on image information from the image sensor. By scanning the object in the optical axis direction and searching for the focal point at all points in the plane, the height distribution of the object can be measured. The focal point is determined using the image contrast distribution obtained by differential processing and so on. The position along the optical axis with the highest contrast is determined to be the focal point, which is the surface of the object to be measured. This method is called the focus variation method or the shape-from-focus method. The diffraction limit defines the lateral resolution, and the resolution in the optical axial depends on the depth of focus. Although the optical axial resolution is worse than the confocal method and white light interferometry described below, the measurement system is simple, so this method can be used for diverse measurement objects. Smooth or mirror-like surfaces pose measurement challenges because obtaining adequate image contrast can be particularly difficult in such cases.

● 9.4　Confocal microscopy

Scanning confocal microscopy can measure microscopic features of the mi-

る．光を測定物に照射すると表面近傍で光は反射・散乱する．共焦点法では，この表面近傍での反射・散乱の光学応答を選択的に取得することで表面の位置を検出する．構成を Fig. 9-7 に示す．光源はレーザを用い，ピンホールなどで点光源とする．この点光源が物体面に投影，つまり焦点面に照射される．また，検出系の像面側共役点にピンホールを配置し，焦点位置で反射や散乱される光のみを取得する．このピンホールを透過した光量を検出器で測定する．このように点光源と点検出（ピンホール）がともに物体面の1点と共役となるように配置するため，共焦点と呼ばれる．測定表面が焦点位置にあるときは多くの光がピンホールを通過し，焦点から外れるとピンホールによって光が遮断される（Fig. 9-7(a)）．光透過可能で限りなく小さい径を持つピン

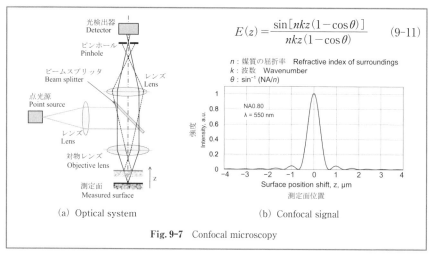

$$E(z) = \frac{\sin[nkz(1-\cos\theta)]}{nkz(1-\cos\theta)} \quad (9\text{-}11)$$

n：媒質の屈折率　Refractive index of surroundings
k：波数　Wavenumber
θ：$\sin^{-1}(\mathrm{NA}/n)$

(a) Optical system　　　　　(b) Confocal signal

Fig. 9-7　Confocal microscopy

cro-scale object. When light is incident onto a measured object, the light is reflected and scattered on the surface. The surface position is detected in the confocal method by selectively acquiring the optical response. The configuration is shown in Fig. 9-7. A laser is used as the light source. A pinhole is set in front of the laser to treat illumination as a point light source. This point light source is projected at the focal point, namely onto the object surface. Another pinhole is placed at the conjugate point on the image side of the detection system. The amount of light transmitted through this pinhole is measured by the detector. The system is called confocal microscopy because the light source and detector are both conjugated to a point on the object plane. When the measured surface is at the focal point, light passes through the pinhole; when at the defocused point, the pinhole blocks the light (Fig. 9-7(a)). By assuming an infinite pinhole, the amplitude $E(z)$ of the transmitted light is given by Eq. (9-11). When scanning the measured target in the optical

ホールを仮定すると，透過する光の振幅 $E(z)$ は式(9-11)で与えられる．測定対象を光軸方向（Z 方向）に走査すると，Fig. 9-7(b)に示す共焦点信号を得る．この信号のピーク位置を表面として検出する．共焦点光学系ではピンホールを配置するため点計測となる．面内の様々な点で測定するためには，多数のピンホールを持つディスク（ニポウディスク）を回転させる，あるいはビーム偏向による 2 次元の走査を行う必要がある．

● 9.5　クロマティック共焦点顕微法

9.2 節で述べたように，レンズは色収差を持ち，異なる波長では光軸方向に広がった焦点位置を持つ．このレンズの色収差特性を利用したものがクロマティック共焦点法である．共焦点顕微法とほとんど同じ構成であるが，光源に広い波長幅を持った白色光あるいは超短パルスレーザ光（第 14 章）を用い，検出器には分光器を用いる（Fig. 9-8(a)）．分光器では，測定物表面の位置に応じて，Fig. 9-7 の共焦点信号に似たスペクトル信号を得る（Fig. 9-8(b)）．この波長信号のピーク波長が測定面位置に対応する（Fig. 9-8(c)）．クロマティック共焦点光学系も点計測であるが，広い波長幅の光を照射し，波長ごとに信号を得るため，測定面の光軸方向（Z 軸方向）走査は不要

axis direction (Z direction), the confocal signal is shown in Fig. 9-7(b). The surface position can be estimated based on the peak position of the confocal signal. In confocal microscopy, only a single point can be measured at one scan due to detection with the pinhole. Therefore, the entire surface is measured by rotating a disk with many pinholes (Nipow disk) or two-dimensional scanning with beam deflection.

● 9.5　Chromatic confocal microscopy

As described in Section 9.2, the lens system has chromatic aberration. Therefore, the focal position of different wavelengths is spread along the optical axis. Chromatic confocal microscopy utilizes this chromatic aberration. An optical system is similar to confocal microscopy, but white light or ultrashort pulse laser (Chapter 14) is used as the light source and a spectrometer as the detector, as shown in Fig. 9-8 (a). The spectrometer obtains a spectral signal similar to the confocal signal in Fig. 9-7, depending on the position of the surface of the measured object (Fig. 9-8 (b)). The peak wavelength of this spectral signal corresponds to the measured surface position (Fig. 9-8(c)). Chromatic confocal optics is also a point measurement. However, it does not require scanning the measured surface in the optical axis (Z-axis) because it obtains signals at each wavelength using chromatic aberration.

となる. 表面の高さ分布を測定するためには, 水平走査は必要となる.

● 9.6 白色干渉法

白色干渉法では, 白色光の短い可干渉距離を応用することで, ナノメートルスケールの極めて高い分解能で表面形状を計測する. 照明に用いる光源の線幅が広い場合, コヒーレンス長が短く, 測定面と参照面の光路差が0に近い場合のみ干渉する (第6章参照). この特性を用いて, 参照面に対する測定面位置を決定する. Fig. 9-9(a)に光学システムを示す. 光源には100 nmオーダの波長幅を持つ白色光源が用いられる. 白

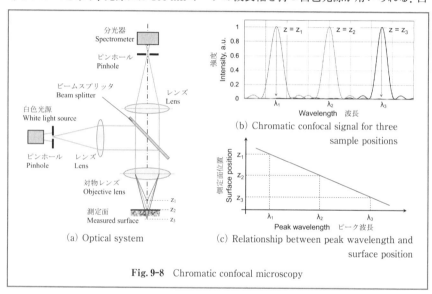

(a) Optical system

(b) Chromatic confocal signal for three sample positions

(c) Relationship between peak wavelength and surface position

Fig. 9-8   Chromatic confocal microscopy

Horizontal scanning is necessary to measure the height distribution of the surface.

● 9.6 White light interferometry

White light interferometry applies the short coherent distance of white light to measure surface topography with high resolution at the sub-nanometer scale. Interference occurs only when the OPD between the measurement and reference surfaces is close to 0 for white light illumination (Chapter 6). This low coherence interference is applied to detect the surface. Fig. 9-9(a) shows the optical system. A white light source with a wavelength range of about 100 nm is used as the light source. The white light interferometer uses a special objective lens with an internal reference plane like the Michelson and Mirau types. Reflected lights from the reference and measured surfaces are interfered with each other on an image sensor

色干渉計では，参照面を内部に組み込んだ特殊な対物レンズを用いる（マイケルソン型，ミロー型）．参照面と測定面からの反射光をイメージセンサ（CCD）上で干渉させる．単色の場合の干渉信号は式(9-12)で表される．k は波数，Δx は参照面（h）と測定面（z）の光路長差（$h-z$），ϕ は反射時の位相シフト量である．白色干渉では，各波長の干渉信号の重ね合わせであるため，式(9-13)のように積算する．$s(k)$ は，光源の持つ波長スペクトルである．測定対象を光軸方向に走査すると，ビームスプリッタを基準とした参照面との光路長差がゼロになる位置でのみ強い干渉を示し，Fig. 9-9(b)のようなインターフェログラムと呼ばれる干渉信号を得る．このインターフェログラ

$$I_k(k)=I_1+I_2+2\sqrt{I_1 I_2}\cos(k\Delta x-\phi), \quad \Delta x=h-z \tag{9-12}$$

$$I=\int s(k)I_k(k)\,dk \tag{9-13}$$

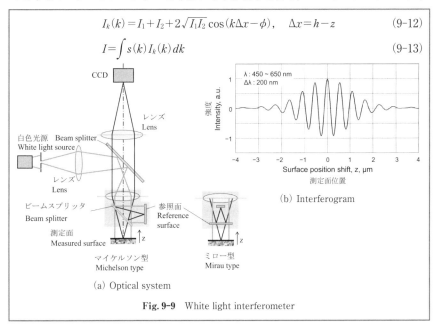

(a) Optical system

(b) Interferogram

Fig. 9-9 White light interferometer

(CCD). The interference signal in the monochromatic case is expressed by Eq.(9-12), where k is the wavenumber, Δx is the optical path length difference ($h-z$) between the reference (h) and measured surfaces (z), and ϕ is the amount of phase shift during reflection. In white light interference, where interference signals $I_k(k)$ occur across various wavelengths, they are integrated as in Eq.(9-13). $s(k)$ is the spectrum of the light source. When the measured object is scanned along the optical axis, strong interference is observed only at the position where the optical path length difference from the reference plane to the beamsplitter is zero. The interference signal I, called an interferogram, is obtained, as shown in Fig. 9-9(b). The optical path length difference is determined by performing envelope and Fourier do-

ムに対し，包絡線解析やフーリエ領域解析を行いピーク位置を検出することで，光路長差を求め測定面高さを求める．イメージセンサで干渉信号を取得できるため面計測可能であり，光軸方向走査のみで表面形状が計測可能である．

【演習問題】

9-1) 式(9-6)の回折限界に関し，レイリー基準の他にスパロー基準なども用いられる．「スパロー基準」の意味とその際用いる α の値を調べよ．

9-2) 式(9-5)では PSF の FWHM を基準とすることから 0.51 が導かれる．$1/e^2$ を用いた場合の係数を求めよ．

9-3) 波長 532 nm で NA 0.6 の対物レンズを用いた際の回折限界と焦点深度を求めよ．回折限界にはレイリー基準を用いるとし，周囲環境は空気の場合を考える．

9-4) 収差補正には組み合わせレンズが用いられる．球面収差を補正するには，どのようなレンズを組み合わせるのかを調べよ．

9-5) 白色干渉計は粗い面や急な斜面の測定ができない．その理由を述べよ．

main analyses on the interferogram and detecting the peak position. Finally, the measurement surface height is obtained. Since the image sensor can acquire interference signals, surface measurement by a single scan is possible. Surface topography can be measured only by scanning in the optical axis direction.

【Problems】

9-1) With respect to the diffraction limit in Eq.(9-6), the Sparrow criterion is used as well as the Rayleigh criterion, investigate the meaning of "the Sparrow criterion" and the value of α to be used in Sparrow criterion.

9-2) Eq.(9-5) leads to 0.51 since the FWHM of PSF is used. Calculate the coefficient when $1/e^2$ is used.

9-3) Calculate the diffraction limit and depth of focus when an objective lens with an NA 0.6 is used at a wavelength of 532 nm.

9-4) Combination lenses are used to correct aberrations. Find out what kind of lens combination is used to correct spherical aberration.

9-5) A white light interferometer cannot measure rough surfaces or steep slopes. State the reason for this.

第 10 章 走査プローブ顕微鏡

走査（型）プローブ顕微鏡（SPM）は，鋭利な探針（プローブ）を試料表面に接近させ，物理的な相互作用を局所的に検出する顕微鏡の総称である．SPM は試料表面をならうようにプローブを走査することで nm オーダの分解能で表面形状や電子状態，機械特性，光学特性などを測定できるため，超精密部品や半導体材料，生体試料の観察や測定に使用されている．本章では，主に表面形状計測について述べる．

● 10.1 走査プローブ顕微鏡の基本

Fig. 10-1 に示すように，表面形状計測に用いる SPM は局所的な物理量を検出する

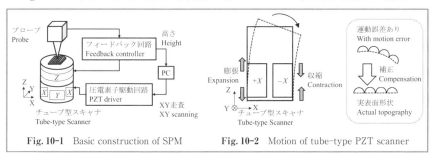

Fig. 10-1　Basic construction of SPM　　　**Fig. 10-2**　Motion of tube-type PZT scanner

Chapter 10　Scanning Probe Microscopy

Scanning probe microscopy (SPM) is a general term for microscopes that can detect local physical interactions by bringing a fine probe close to the sample surface. SPM is widely used for observation and measurement of ultraprecision parts, semiconductor devices, and biological samples because the SPM can measure the surface profile, electronic properties, mechanical characteristics, optical characteristics, and other properties at nanometer-scale resolution. This chapter mainly describes the topography measurement of the sample surface with SPMs.

● 10.1 Fundamentals of SPM

As shown in Fig. 10-1, SPMs used for surface topography measurement comprise a probe that can detect local physical quantities, a feedback controller that controls

プローブ，プローブ先端–試料表面間距離を制御するフィードバック回路，測定点の位置決めや走査を行うスキャナから構成される．プローブは試料表面との距離により変化する局所的な相互作用の物理量，例えば，トンネル電流や分子間力，静電気力などを検出するため，プローブ先端の曲率半径は nm オーダに先鋭化される．測定中のプローブ–試料間距離は，プローブが検出する局所的な相互作用の物理量が一定となるようフィードバック回路を介してスキャナの Z 軸変位を制御することで一定に保持される．プローブ–試料間距離を一定に制御した状態で XY 方向に変位すると，プローブ先端は試料表面をならうように試料上を走査する．走査中のスキャナの Z 軸変位を集積し，三次元表示することで試料の表面形状が取得できる．SPM のスキャナには nm オーダの位置決め分解能とサブ mm〜mm オーダの駆動範囲が要求され，XYZ 各軸に伸縮可能なチューブ型圧電スキャナが広く用いられる．パーソナルコンピュータ（PC）より出力された走査信号を圧電素子駆動回路で増幅し，スキャナ XY 各軸の圧電素子に印加することで走査が行われる．また，チューブ型圧電スキャナ先端は Fig. 10-2 に示すように円弧運動するため，プローブ走査により得られた表面形状よりスキャナ運

the distance between the probe tip and the sample surface, and a scanner that positions and scans the measurement points. Since the probes detect physical quantities of local interactions that change with the distance from the sample surface, such as tunnel current, intermolecular force, and electrostatic force, the curvature radius at the probe tips is sharp on the order of nm. The probe–sample distance during the measurement is maintained constant by controlling the Z-axial displacement of the scanner via the feedback controller so that the physical quantity of local interaction detected by the probe remains constant. When the scanner is moved in the XY direction while the probe–sample distance remains constant, the probe tip scans the sample to follow the sample surface. The topography of the sample surface can be obtained by accumulating the Z-directional scanner displacement during scanning and displaying it in three dimensions. The scanners for SPMs are required to realize a positioning resolution in the order of nm and strokes of sub-mm to mm order. Therefore, tube-type piezoelectric actuators that can move in each XYZ axis are widely utilized. Two-dimensional scanning is conducted by amplifying the scanning signal output from a personal computer using PZT drivers and applying it to the PZT actuators of the scanner in the XY axes. In addition, since the top of the tube-type PZT scanner moves in an arc as shown in Fig. 10-2, a more accurate surface topography can be obtained by correcting the scanner motion error from the sur-

動誤差を補正することによってより正確な表面形状が取得できる.

● 10.2   走査トンネル顕微鏡

走査（型）トンネル顕微鏡（STM）は先鋭化した導電性プローブと試料との間を流れるトンネル電流を検出し，非接触で表面形状を計測する．Fig. 10-3 に示すように，1 nm 未満の距離まで接近させた非接触状態のプローブ–試料間にバイアス電圧を印加すると，電子はトンネル効果によってプローブの高いフェルミ面から試料の低いフェルミ面に移動する．プローブと試料の仕事関数をそれぞれ ϕ_p および ϕ_s とすると，トンネル電流 I とプローブ–試料間距離 z の関係は式(10-1)となり，トンネル電流は指数関数的に変化する．ここで m は電子質量，$\hbar$ は換算プランク定数である．プローブ–

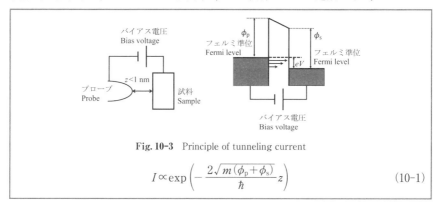

Fig. 10-3   Principle of tunneling current

$$I \propto \exp\left(-\frac{2\sqrt{m(\phi_\mathrm{p}+\phi_\mathrm{s})}}{\hbar}z\right) \tag{10-1}$$

face topography measured by the probe scanning.

● 10.2   Scanning tunneling microscopy

Scanning tunneling microscopy (STM) is a noncontact measurement method of surface topography that detects the tunneling current between the sharpened conductive probe and the sample. As shown in Fig. 10-3, when a bias voltage is applied with the distance between the probe and sample less than 1 nm, electrons are tunneled from the higher Fermi surface of the probe to the lower Fermi surface of the sample. Here, the work function of the probe and sample are defined as ϕ_p and ϕ_s, respectively. The relationship between tunneling current I and probe–surface distance z is given by Eq. (10-1), and the tunneling current varies exponentially. Here, m and $\hbar$ are the electron mass and the reduced Planck constant, respectively. As the probe–sample distance increases, the probability of tunneling decreases significantly, indicating that the tunneling current flows between the nearest atoms of the probe and sample. Consequently, the STM can realize high vertical resolution. How-

試料間距離が長くなると，トンネル現象の確率は著しく減少する．結果として，トンネル電流はプローブと試料の最も近い原子間に流れ，STM は高い垂直分解能を実現できる．一方，水平分解能はプローブ先端の形状や曲率半径に影響され，垂直方向と比較し低下する．Fig. 10-4 に STM の構成を示す．機械研磨や電解研磨により先鋭化した Pt や Pt/Ir，Ni，W などが STM プローブとして使用される．プローブ-試料間を流れるトンネル電流は高感度電流-電圧変換回路により検出され，対数アンプを介してフィードバック回路に入力される．

STM の測定方法には，スキャナの Z 方向変位を固定してトンネル電流の変化から形状を直接算出する Constant-height mode（CHM）と，トンネル電流が一定となるようにスキャナ Z 軸変位を制御する Constant-current mode（CCM）がある．CHM は迅速な測定が可能であるが，Z 方向測定範囲が狭い．CCM では測定時間は増加する

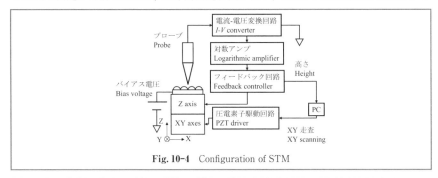

Fig. 10-4 Configuration of STM

ever, the shape and curvature radius of the probe tip affect the horizontal resolution, making it lower than the vertical resolution. Fig. 10-4 shows the typical configuration of STM. Pt, Pt/Ir, Ni, W, and other conductive materials sharpened by mechanical or electrochemical polishing are used as STM probes. The tunneling current flowing between the probe and sample is detected using a highly sensitive current-to-voltage converter and detected via a logarithmic amplifier.

The measurement methods of the STM include constant-height mode (CHM), in which the Z-direction displacement of the scanner is fixed and the topography is calculated directly from changes in the tunneling current, and constant-current mode (CCM), in which the Z-direction displacement of the scanner is feedback controlled, so that the tunneling current remains constant. Although the CHM is capable of rapid measurement, the measurement range along the Z-direction is restricted. In contrast, the CCM increases the measurement time but, it can provides both higher resolution and a wider measurement range in the Z-direction. The vertical and horizontal resolutions of the STM are approximately 0.01 nm and 0.1 nm,

が，Z 方向の高分解能と広測定範囲を両立させる．STM の垂直分解能と水平分解能は
それぞれ 0.01 nm および 0.1 nm 程度であり，原子レベルの構造や電子状態を観察で
きる．一方，測定対象は導電性試料に限られる．大気中ではプローブや試料表面の酸
化被膜の影響を受けるため，真空中や超高真空中での測定が行われている．

● 10.3 原子間力顕微鏡

原子間力顕微鏡（AFM）は，カンチレバー（片持ち梁）先端付近に取り付けた鋭利
なプローブと試料表面の接触や力学的な相互作用を検出し，表面形状を測定する．
AFM 測定における代表的な力学モデルとして Fig. 10-5 に示すレナード–ジョーン
ズ・ポテンシャルがある．2 個の無極性原子間には式(10-2)で近似される相互作用力
が作用する．ここで，z は 2 原子間距離，ε と σ はフィッティングパラメータである．

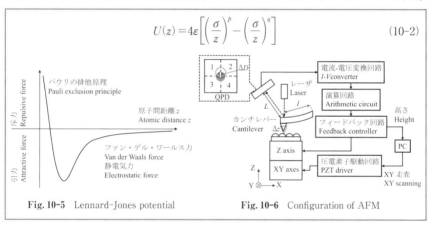

$$U(z) = 4\varepsilon\left[\left(\frac{\sigma}{z}\right)^{p} - \left(\frac{\sigma}{z}\right)^{q}\right] \tag{10-2}$$

Fig. 10-5 Lennard-Jones potential   **Fig. 10-6** Configuration of AFM

respectively. While STM facilitates the observation of atomic-level structures and
electronic states, its applicability is confined to conductive materials. Moreover,
measurements are conducted within a vacuum or ultrahigh vacuum environment
due to the impact of surface oxide films on the probe and the sample when exposed
to atmospheric air.

● 10.3 Atomic force microscopy

Atomic force microscopy (AFM) can detect the contact or local physical interaction between the probe and the sample surface with a sharpened tip attached near
the edge of the cantilever beam, and the surface topography can be measured. As a
typical dynamic model in AFM measurement, a schematic of the Lennard-Jones
potential is shown in Fig. 10-5. An interaction force approximated by Eq. (10-2)
applies between two nonpolar atoms, where z is interatomic distance, ε and σ are the

q 乗項は引力項を示し，2 原子が離れている場合，ファン・デル・ワールス力や静電気力による引力が作用する．p 乗項は斥力項を示し，パウリの排他原理による斥力の相互作用を意味している．Fig. 10-6 は光てこ法を用いた AFM の構成を示す．AFM はプローブ–試料間の相互作用力をカンチレバーのたわみとして検出する．カンチレバー背面の先端にレーザを照射し，その反射光を 4 分割フォトダイオード（QPD）で検出する．プローブに相互作用力が働くと，カンチレバーにたわみが生じ，QPD 上のスポット位置が変化する．カンチレバーの長さを l，カンチレバーと QPD の距離を L，カンチレバー先端変位を Δz とすると，QPD 上でのスポット変位 ΔD は式(10-3)のように拡大される．QPD の各受光面の出力電圧 V_1〜V_4 に式(10-4)を適用して検出される ΔV_{diff} はカンチレバーのたわみによるレーザスポットの上下方向変位に相当する．

Fig. 10-7(a)に示すカンチレバーの静的なたわみを検出する方法はコンタクトモード AFM と呼ばれる．コンタクトモード AFM はプローブ–試料間の相互作用力をカンチレバーのたわみ ΔD として検出し，たわみが一定の設定値 ΔD_{set} に維持されるよう

$$\frac{\Delta D}{\Delta z} = \frac{2L}{l} \tag{10-3}$$

$$\Delta V_{\mathrm{diff}} = (V_1 + V_2) - (V_3 + V_4) \tag{10-4}$$

fitting parameters. The q-power term indicates the attractive force term, and when two atoms are separated, the attractive force derived from the van der Waals force or electrostatic force is applied. The p-power term indicates the repulsive force term, which means the interaction of the repulsive force due to the Pauli exclusion principle. Fig. 10-6 shows the configuration of AFM using an optical lever system. AFM detects the interaction force applied to the probe tip as the deflection of the cantilever. The laser beam is irradiated to the back of the cantilever, and the reflected light is detected by a quadrant photodiode (QPD). When interaction forces act on the probe tip, the cantilever deflects and the laser spot position on the QPD is moved. The spot displacement ΔD on the QPD is magnified, as shown in Eq. (10-3). Here, l is the length of the cantilever, L is the distance between the cantilever and the QPD, and Δz is the displacement of the cantilever. ΔV_{diff}, which is detected by applying Eq. (10-4) to the output voltages V_1 to V_4 of each channel of the QPD, corresponds to the vertical displacement of the laser spot due to the deflection of the cantilever.

As shown in Fig. 10-7(a), the method that detects the static deflection of the cantilever is called contact-mode AFM. In contact-mode AFM, the interaction force between the probe tip and the sample surface is detected as the deflection ΔD of the cantilever. The surface topography of the sample is measured by scanning

スキャナの Z 方向変位を制御した状態で走査し，試料の表面形状が測定される．コン
タクトモード AFM の測定力はカンチレバーのばね定数とたわみから算出できる．よ
り低測定力の方法として，Fig. 10-7(b)に示す振動プローブを用いたダイナミック
モード AFM が開発された．共振周波数近傍で Z 方向に加振したカンチレバーを試料
表面に接近させると，試料表面との相互作用力や断続的な接触によってカンチレバー
の振動状態に変化が生じる．このときのカンチレバー振動の振幅 A や共振周波数シフ
ト Δf はプローブ-試料間距離によって変化するため，これらを一定に保持するように
スキャナ Z 方向変位を制御することでプローブ-試料間距離が一定に保持される．
Fig. 10-7(c)のように，振幅 A を一定の設定値 A_{set} に保持する方法は振幅変調（AM）

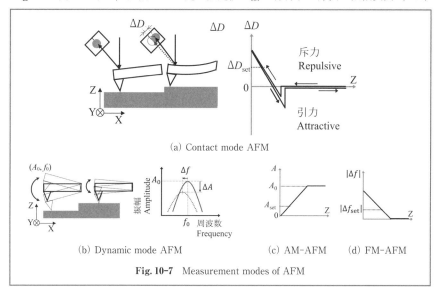

(a) Contact mode AFM

(b) Dynamic mode AFM   (c) AM-AFM   (d) FM-AFM

Fig. 10-7 Measurement modes of AFM

while controlling the Z-directional displacement of the scanner so that ΔD is maintained at a constant set value of ΔD_{set}. The measurement force can be calculated
based on the spring constant of the cantilever and deflection. The dynamic-mode
AFM, which utilizes a vibrating probe, as depicted in Fig. 10-7(b), has been developed as a method with reduced measuring force. When a cantilever vibrating on its
resonance frequency in the Z-direction is brought close to the sample surface, the
vibration condition is changed due to the interaction forces or intermittent contact
with the surface. Because the amplitude A or the frequency shift amount Δf of the
probe vibration is varied by the probe-to-surface distance, the changes in the
probe vibration characteristics can be maintained constant by controlling the Z-directional displacement of the scanner to keep them constant. As shown in Fig. 10-7

AFM と呼ばれ，カンチレバーの振動振幅信号が制御回路に入力される．Fig. 10-7(d) のように，相互作用力により生じるカンチレバーの共振周波数シフト $|\Delta f|$ を一定の設定値 $|\Delta f_{\mathrm{set}}|$ に保持する方法は周波数変調（FM）AFM と呼ばれ，位相同期（PLL）回路を用いて共振周波数シフトを検出し，制御回路に入力する．FM-AFM は AM-AFM と比較して高い力検出の感度と応答性を実現できる．AFM プローブは一般にシリコンや窒化シリコンから半導体加工技術を応用して製作される．AFM は試料の導電性に影響されないため，絶縁体や有機材料，生体試料の測定にも広く用いられ，真空中以外にも大気中や溶液中でも動作可能である．

● 10.4　走査静電気力顕微鏡

STM や AFM は局所的な相互作用を検出するためにプローブ-試料間距離を 1 nm 未満に保持する必要があり，表面凹凸が大きい試料の測定では測定時間の増加やプローブ衝突のリスクが課題となる．走査静電気力顕微鏡（SEFM）は，STM や AFM より大きなプローブ-試料間距離で試料表面形状を非接触測定するために静電気力を検出する．Fig. 10-8 に SEFM の構成を示す．SEFM は金属プローブを使用し，プロー

(c), the method to maintain amplitude A to a constant set value A_{set} is called amplitude modulation (AM) AFM, and the vibration amplitude signal of the cantilever is fed into the control circuit. As shown in Fig. 10-7(d), the method to maintain the resonance frequency shift $|\Delta f|$ of the cantilever vibration due to the interaction forces to a constant set value $|\Delta f_{\mathrm{set}}|$ is called frequency modulation (FM) AFM, and the resonance frequency shift signal, which is detected using a PLL circuit, is input in the control circuit. FM-AFM can achieve higher sensitivity for interaction force detection and responsibility than AM-AFM. Since AFM does not affect sample conductivity, it is also widely used for measuring insulating materials, organic materials, and biological samples. In addition, AFM can operate not only in a vacuum but also in the atmosphere and solutions.

● 10.4　Scanning electrostatic force microscopy

In STM and AFM, the pro-sample distance should be maintained at less than 1 nm to detect local interactions. Therefore, in the measurement of samples with large surface unevenness, such as diffraction gratings, they are time consuming, and there is a risk of probe collision. Scanning electrostatic force microscopy (SEFM) detects a probe-sample distance larger than those of STM or AFM as electrostatic force for noncontact measurement of the topography. Fig. 10-8 shows the configura-

ブ–試料間に印加したバイアス電圧の電場によってプローブ先端–試料表面間には静電気による引力が生じる．このとき，任意の測定点 x においてプローブ先端に作用する静電気力 $F_E(x)$ の大きさとプローブ–試料間距離 $h(x)$ の関係は式(10-5)となる．Z 方向に共振周波数振動させたプローブに非接触状態で引力型の静電気力が作用すると，プローブの共振周波数に変化が生じる．静電気力によるプローブ共振周波数シフト Δf (x) は，プローブ振動系のばね定数を k とすると式(10-6)で示される．ここで，ε_0 と ε_γ は真空の誘電率および雰囲気の比誘電率，R はプローブ先端曲率半径，V_{dc} はバイアス電圧，$V_{cpd}(x)$ は接触電位差である．プローブ共振周波数の変化は前述の FM-AFM と同様に PLL 回路を用いて検出される．一方，試料表面の電荷の不均一やプロー

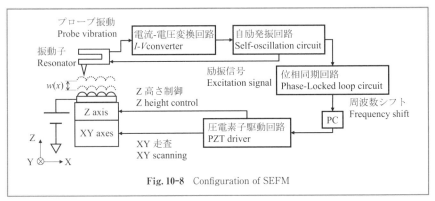

Fig. 10-8 Configuration of SEFM

tion of the SEFM. SEFM employs a metal probe. Electrostatic force is generated at the probe tip due to the electric field of the bias voltage applied between the probe and the sample. The relationship between the intensity of the electrostatic force $F_E(x)$ acting on the probe tip at an arbitrary measurement point x and the distance $h(x)$ between the probe and sample is given by Eq. (10-5). When an attractive electrostatic force is applied on a probe vibrated at the resonance frequency in the Z direction without contact, the resonance frequency of the probe changes. The resonance frequency shift of the probe vibration $\Delta f(x)$ due to electrostatic force can be expressed by Eq. (10-6), where k is the spring constant of the probe vibration system. ε_0 and ε_γ represent the permittivities of vacuum and atmosphere, respectively. R is the curvature radius of the probe tip, V_{dc} is the bias voltage, and $V_{cpd}(x)$ is the contact potential difference. The variation in the probe resonance frequency is detected using a PLL circuit as in the FM-AFM described above. However, the frequency shift does not coincide with the topography due to the probe tip shape or the inhomogeneous charge on the sample surface. Therefore, the probe-surface distance is changed by $w(x)$, and the same profile is rescanned. The resonance fre-

ブ先端形状によって周波数シフトと試料表面形状は一致しない．SEFM はプローブ-試料間距離を $w(x)$ 変化させ，試料上の同一箇所を再度走査する．プローブ-試料間距離を $w(x)$ 変化したときのプローブ振動の共振周波数シフト $\Delta f_w(x)$ は式(10-7)となる．$w(x)$ が十分に小さいとき，式(10-6)と式(10-7)を連立し，$h(x)$ について解くと式(10-8)に示すように，誘電率や接触電位差の影響が相殺され，プローブ振動周波数シフトから算出できる．SEFM のプローブには高い振動の Q 値が要求され，音叉型水晶振動子の梁に先鋭化した金属線を取り付けてプローブとして使用される．水晶振動子は圧電効果により歪みに応じた電流を発生させることから，光てこ光学系が不要な小型でシンプルな構成となる．

$$F_E(x) = \pi\varepsilon_0\varepsilon_\gamma \frac{R}{h(x)} (V_{dc} + V_{cpd}(x))^2 \tag{10-5}$$

$$\Delta f(x) = \frac{f}{2k}\frac{\partial F}{\partial h} = -\frac{f_0}{2k}\pi\varepsilon_0\varepsilon_\gamma(V_{dc} + V_{cpd}(x))^2\frac{R}{h^2(x)} \tag{10-6}$$

$$\Delta f_w(x) = -\frac{f_0}{2k}\pi\varepsilon_0\varepsilon_\gamma(V_{dc} + V_{cpd}(x))^2\frac{R}{(h(x)+w(x))^2} \tag{10-7}$$

$$h(x) = w(x)\frac{\sqrt{\Delta f_w(x)/\Delta f(x)}}{1-\sqrt{\Delta f_w(x)/\Delta f(x)}} \tag{10-8}$$

quency shift $\Delta f_w(x)$ of the probe vibration when the probe-sample distance changes $w(x)$ is expressed by Eq. (10-7). When $w(x)$ is sufficiently small, solving Eqs. (10-6) and (10-7) for $h(x)$ cancels out the influence of the dielectric constant and contact potential difference, as shown in Eq. (10-8). Consequently, the topography can be calculated based on the frequency shift of the probe vibration. The SEFM probes require a high Q-factor of the vibration, and a tuning-fork quartz crystal resonator with a sharpened metal wire attached to the beam is used as the probe. Since the quartz crystal generates a current according to the strain owing to the piezoelectric effect, it allows a compact and simple probe configuration that does not require an optical lever optical system.

【演習問題】

10-1) STM による表面形状の測定以外の応用を述べよ.

10-2) 式(10-2)において, $p=12$, $q=6$ の場合, 引力が最大となる z を求めよ.

10-3) SPM により測定可能な表面形状とプローブ形状の関係について示せ.

【Problems】

10-1) Describe the applications of STMs other than surface topography measurement.

10-2) In Eq. (10-2), when $p=12$ and $q=6$, find z that maximizes the attractive force.

10-3) Show the relationship between the surface topography measurable by SPMs and the probe tip form.

第11章　誤差要因と不確かさ

　計測における誤差は「測定の結果から測定対象量の真の値を引いたもの」(GUM)
と定義されるが，実際の測定では真の値を具体的に知ることはできない．このため，
誤差の概念では不可知量を扱う困難を生ずる．それに対し，真値を知ることができな
い前提で測定値のばらつきを表現するのが不確かさの概念である．不確かさは「測定
値に付随する，合理的に測定対象量に結び付けられうる値のばらつきを特徴付けるパ
ラメータ」と定義され，厳密な測定においては測定値に不確かさを付記して完全なも
のとなる．不確かさ評価は GUM に従って行われる．不確かさ要因は測定対象に起因
するもの，測定機器に起因するもの，環境に起因するものなど，すべての要因につい
て検討し，そのうち影響の大きなものをピックアップして不確かさのバジェットシー
トを作り評価を行う．バジェットシートの「測定量の単位で表記した標準不確かさ」
の項から各不確かさ要因の寄与割合が明らかになり，また，事前の検討や改善効果の

Chapter 11　Error Sources and Measurement Uncertainty

　Error of measurement is defined as the "result of a measurement minus a true
value of the measurand" in GUM (Guide to the Expression of Uncertainty in Measurement). However, since the true value is by nature indeterminate except for
some fundamental physical constants, the exact error of a measurement result, in
general, cannot be obtained. The concept of measurement uncertainty is to express
the dispersion of the measured values based on this fact. Measurement uncertainty
is defined as the "parameter, associated with the result of a measurement, which
characterizes the dispersion of the values that could reasonably be attributed to the
measurand." It is reported with a measurement result. Uncertainty analysis is performed based on GUM. In this analysis, all the possible sources of uncertainty, such
those from the measurand, the measuring instrument, and the environment, should
be considered. An uncertainty budget is then made to evaluate uncertainty by selecting the major sources of uncertainty. The impact of each source of uncertainty
can be classified from the standard uncertainty with the unit of the measurand in

見積りが可能となる点からも不確かさ評価は有効である.

　従来は系統的誤差として扱っていた偏りも系統的効果による不確かさとして扱いうるが，校正可能な場合や，合理的に補正可能な場合にはそれらをすべて行った後に残るばらつきに対して不確かさを算出する．本章ではそのような系統誤差のうち，精密工学分野でよく見られる誤差要因について説明する．なお，不確かさの求め方については本書の姉妹編『Bilingual edition 計測工学』（朝倉書店）を参照されたい.

● 11.1　アッベ誤差

　Fig. 11-1 にスケール（ものさし）を使って物体の長さを測る簡単な例を示す．物体の辺 AB の長さをスケールで精度よく測るには，スケールを辺 AB に当てて，それぞれ A 点と B 点の真上からスケールを読み取る必要がある．A, B 点のスケールの読み

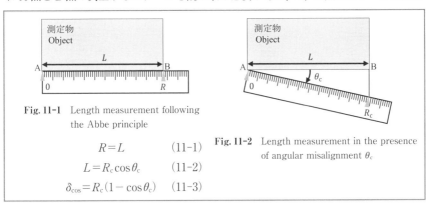

Fig. 11-1　Length measurement following the Abbe principle

$$R = L \qquad (11\text{-}1)$$

$$L = R_\mathrm{c} \cos \theta_\mathrm{c} \qquad (11\text{-}2)$$

$$\delta_\mathrm{cos} = R_\mathrm{c}(1 - \cos \theta_\mathrm{c}) \qquad (11\text{-}3)$$

Fig. 11-2　Length measurement in the presence of angular misalignment θ_c

the uncertainty budget. Uncertainty analysis is also effective for planning and improving the measurement.

Bias, which has conventionally been treated as a systematic error, can be treated as a systematic effect-induced uncertainty. However, the uncertainty should be evaluated to characterize the residual dispersion after all possible corrections are made if calibration or reasonable corrections are possible. In this chapter, some major systematic errors in precision engineering are presented. For uncertainty analysis, it is suggested to read "Measurement and Instrumentation" (Bilingual edition, Asakura Publishing).

● 11.1　Abbe errors

Fig. 11-1 shows a length measurement using a scale ruler. To accurately measure the length L of side AB of an object, it is necessary to align the scale ruler to side AB and read the scale graduations from right above points A and B to obtain the

値をそれぞれ 0, R とすると, R は辺 AB の長さ L と一致する（式(11-1)）. このように, 長さを正しく測るためには, 測定対象物と長さ測定機器の測定軸を一直線上に設置しなければならない. これをアッベの原理といい, 長さ計測における最も基本的な原理原則である. アッベの原理を満たす長さ測定機器としてマイクロメータが挙げられる.

一方, 長さ測定機器では, 種々の制約を受けてアッベの原理を満たさない設置条件で長さ計測を行わなければならないケースも多く, 長さ測定の結果に誤差が生じる. このような誤差をアッベ誤差と呼ぶ. アッベ誤差は, 主に長さ測定機器の測定軸と測定対象物の間に存在する

- ミスアライメント角度
- オフセット（アッベオフセットと呼ぶ）

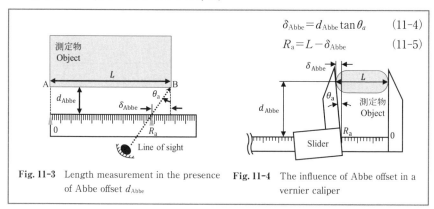

$$\delta_{Abbe} = d_{Abbe} \tan \theta_a \qquad (11\text{-}4)$$

$$R_a = L - \delta_{Abbe} \qquad (11\text{-}5)$$

Fig. 11-3　Length measurement in the presence of Abbe offset d_{Abbe}

Fig. 11-4　The influence of Abbe offset in a vernier caliper

readings of 0 and R, respectively. In this case, R is equal to the length L (Eq. (11-1)), where the measurement follows the Abbe principle of alignment, the most fundamental principle in length measurement. The principle states that the maximum accuracy of length measurement is achieved when the axes of the length-measuring instrument and the object being measured are aligned coaxially. The micrometer is one of the measuring instruments satisfying the Abbe principle.

However, there are many instances where the length-measuring instrument fails to satisfy the Abbe principle. In such cases, errors will occur in the measurement results. This error is called the Abbe error. Abbe errors are primarily caused by the following factors of misalignment between the axes of the measuring instrument and the object being measured.

- misalignment angle
- offset (Abbe offset)

によって生じる.

　Fig. 11-2 にミスアライメントによってスケールが θ_c 傾斜していた場合を示す. 実寸法 L と B 点の読み値 R_c の関係は式(11-2)のようになり, 式(11-3)に示す測定誤差 δ_{cos} が生じてしまう. 一般にこの誤差成分をコサイン誤差と呼ぶ. この誤差は変位測定の場合にも移動軸と変位測定軸のミスアライメント角により発生する.

　一方, アッベオフセット d_{Abbe} が存在する場合を Fig. 11-3 に示す. このとき, スケールを読み取る視線の方向が B 点に正対する方向から θ_a だけ偏位すると, 式(11-4)のように測定誤差 δ_{Abbe} が生じ, スケールの読み値が式(11-5)になる. θ_a が十分に小さい場合, δ_{Abbe} は $d_{Abbe}\theta_a$ で表せる. Fig. 11-4 に示すノギスの場合, 測定対象物とスケールの目盛の間にアッベオフセットがある構造となっている. スライダの移動機構にはわずかな遊びが必要であるため, 測定対象物を挟み込む力によりスライダが θ_a 傾斜し, それに伴い測定誤差 δ_{Abbe} が生じる. 一般にアッベ誤差と呼ぶ場合はこの誤差成分を指すことが多い.

　ほとんどの長さあるいは変位測定機器では, δ_{cos} と δ_{Abbe} の両方が存在する. 三次元座標測定機はその典型的な例である. そのため, 三次元座標測定機ではプローブの空間位置校正を行い, その結果をもとに補正を行って精度を確保している.

　Fig. 11-2 shows the case with a misalignment angle θ_c. The scale reading at Point B becomes R_c with an error δ_{cos} from the actual length L as shown in Eqs. (11-2) and (11-3). δ_{cos} is often called the cosine error, which also occurs in displacement measurement when angular misalignment exists between the axes of the measuring instrument and the moving object.

　Fig. 11-3 shows the case with an Abbe offset d_{Abbe}. If the line of sight has an angular deviation θ_a, an error δ_{Abbe} shown in Eq. (11-4) will be generated, and the scale reading at Point B will change to R_a shown in Eq. (11-5). δ_{Abbe} can be expressed as $d_{Abbe}\theta_a$ when θ_a is sufficiently small. As shown in Fig. 11-4, an Abbe offset d_{Abbe} exists between the scale and the object being measured with a vernier caliper. The slide will be titled at an angle θ_a when a force acts on the object due to the clearance between the slider and the main scale, under which the error δ_{Abbe} occurs. This error component is often referred to as the Abbe error.

　Most length- and displacement-measuring instruments, such as a three-dimensional CMM, have the error components of δ_{cos} and δ_{Abbe}. In a CMM, the positions of the probe in space are calibrated so that measurement accuracy can be assured through compensation based on the calibration results.

● 11.2　熱変形

精密測定においては温度変化の影響は測定対象の熱変形を引き起こすだけでなく，測定機器にもおよび，電子機器の特性変化などの形でも測定精度に影響する．ここでは長さ測定における熱膨張の影響について検討する．ISO 1 では製品の幾何的特性仕様の標準基準温度を 20℃ と定めている．しかし，常に 20℃ の環境での測定を行うことはできないため，スケールや測定対象の熱膨張による誤差が生ずる．

Fig. 11-5 に示すように，熱膨張係数 $\alpha_{\mathrm{w}}\,[\mathrm{K}^{-1}]$ の材質で作られた長さ L（すなわち，20℃における長さ L）の物体を 20℃ で値付けられたスケールで測定することを考える．A 点のスケールの読み値を常に 0 とし，20℃ のときの B 点のスケールの読み値を R とする．物体の温度が T_{w} に変化した場合，熱膨張による変化後の長さ L' は式(11-6)で表される．スケールの材質の熱膨張係数を $\alpha_{\mathrm{s}}\,[\mathrm{K}^{-1}]$，スケールの温度を T_{s} とし，熱膨張後の物体の B′ 点が対応するスケールの読み値を R' とすると，式(11-7)が得られ

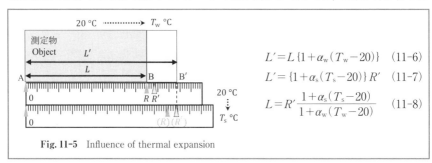

$$L' = L\{1 + \alpha_{\mathrm{w}}(T_{\mathrm{w}} - 20)\} \qquad (11\text{-}6)$$

$$L' = \{1 + \alpha_{\mathrm{s}}(T_{\mathrm{s}} - 20)\}R' \qquad (11\text{-}7)$$

$$L = R'\,\frac{1 + \alpha_{\mathrm{s}}(T_{\mathrm{s}} - 20)}{1 + \alpha_{\mathrm{w}}(T_{\mathrm{w}} - 20)} \qquad (11\text{-}8)$$

Fig. 11-5　Influence of thermal expansion

● 11.2　Thermal deformation

The temperature changes cause not only the thermal deformation of the object being measured but also the performance changes in the electronics of the measuring instruments. Here, the influence of thermal expansion on length measurement is discussed. The standard temperature for the specifications of geometric characteristics is set to be 20℃ in ISO 1. However, since the measurement is not always made at 20℃, measurement errors are caused by the thermal expansion of the scales and the object being measured.

Fig. 11-5 shows an object with a length of L at 20℃. Assume the reading of the scale at point A is always 0. The reading at point B is R when both the object and the scale are at 20℃. The coefficients of thermal expansion of the object and the scale are $\alpha_{\mathrm{w}}\,[\mathrm{K}^{-1}]$ and $\alpha_{\mathrm{s}}\,[\mathrm{K}^{-1}]$, respectively. Owing to the effect of thermal expansion, when the temperatures of the object and the scale change to T_{w} and T_{s}, respectively, the length of the object changes to L' shown in Eq. (11-6) and the reading R' of the scale at point B' has a relationship with L' expressed in Eq. (11-7).

る．式(11-8)より，物体の20℃ときの長さ *L* はスケールの読み値 *R'* から計算できる．

　ここで，物体とスケールを同じ環境に十分になじませ，両者の温度が同じとなった場合を考える．測定物とスケールが同じ材質で作られ，熱膨張係数が同じ場合には *L* ＝*R'*＝*R* となり，温度変化により物体とスケールの実際の寸法は変化するが，スケールの読み値としては20℃における長さを示す．機械部品には鋼が広く用いられるが，鋼製部品を鋼のブロックゲージを基準に検査する場合がこれに相当する．

　また，スケールの熱膨張係数が0とみなせる場合には20℃から多少の温度変動が起こった場合にも誤差が小さく，温度 *T_w* における物体の長さ *L'* をスケールの読み値 *R'* から補正無しに測定できる．長さ測定機器の校正などに用いられる標準尺は熱膨張係数が低い材料で製作され，熱膨張係数が0.02×10^{-6}[K^{-1}] の低熱膨張ガラスが採用されている例がある．

　測定者の手から測定物やスケールに熱が伝わる場合など，場所により温度が異なるケースでは単純に補正式を適用できない．100 mm 長さの鋼製ゲージブロックを10分間素手で扱った場合におよそ8 μm の伸びが生じ，その影響がサブマイクロメートルレベルまで下がるのに60分程度を要するというデータもある．精密測定においては極

The length *L* of the object at 20℃ can then be calculated from *R'* as shown in Eq. (11-8).

When the temperatures of both the object and the scale are the same, *L*, *R'*, and *R* will be identical if the object and the scale are made of the same material, i.e., with the same coefficient of thermal expansion. In this case, although the lengths of the object and the scale change with temperature variation, the scale reading demonstrates the length of the object at 20℃. This is the case when a part made of steel is measured using a steel gauge block as a reference.

If the scale has a sufficiently small coefficient of thermal expansion, the changes in scale readings can be ignored when the temperature deviates slightly from 20℃. The length *L'* of the object at temperature *T_w* can then be obtained from the scale reading *R'* without compensation. The line scales for calibration of length instruments are made of low thermal-expansion materials, such as glasses with a coefficient of thermal expansion of 0.02×10^{-6} [K^{-1}].

When variations in the temperature of an object or scale are induced by the operator's hand, compensating for the measurement results in terms of length becomes challenging due to the uneven distribution of temperature. A steel gauge block of 100 mm long can have an 8 μm expansion when it is handled by hand over 10 min, which takes approximately 60 min for the expansion to decrease to a submi-

力熱を伝えないように細心の注意を払い，温度が安定してから測定する必要があることがわかる．

● 11.3　力による変形

精密さを要求する領域では測定対象や測定基準，測定器の構造のいずれもが剛体とはみなせず，自重や測定力，その他にも荷重や締結力による構造部材の変形を前提に考える必要がある．この変形はフックの法則などを基本とする材料力学により求めることができる．

Fig. 11-6 に示すようにヤング率 E の材質でできた幅 u，厚さ v，長さ L の物体（直方体）の弾性変形について考える．Fig. 11-6 のように，鉛直姿勢で支持されている物体では外部荷重 F の他，物体自重も弾性変形の要因となる．荷重 F による鉛直方向の変形量 ΔL_F は断面方向の変形を無視すると式(11-9)のように近似できる．式(11-9)からわかるように，変形量 ΔL_F は長さ L，荷重 F に比例し，物体のヤング率および断面積 $S(=uv)$ に反比例する．次に物体の自重による弾性変形量を求めるために，物体の下端から z の位置にある長さ dz の微小物体部分の重さを δW_g とし，それによる下部の物体部分に与える弾性変形量を δl_g とする．重力加速度が g，物体の体積密度が ρ の

crometer level. In precision metrology, careful attention to thermal deformations and delaying measurements until the temperature stabilizes are recommended.

● 11.3　Force-induced deformation

When precision is required, the objects, measuring instruments, and measurement standards cannot be treated as rigid bodies. The deformations caused by self-weights, measuring forces and loads should be considered according to the mechanics of materials based on Hooke's low.

Consider the elastic deformations of a vertically supported rectangular object with a width u, a thickness v, a length L, and a Young's modulus E as shown in Fig. 11-6, which can be caused by the external force F and the self-weight of the object. The vertical deformation ΔL_F caused by F can be approximated by Eq. (11-9), which is proportional to L and F and inversely proportional to E and the cross-sectional area $S(=uv)$. To evaluate the vertical deformation caused by the self-weight of the object, denote the weight of the small part with a length dz, at the position z from the bottom, by δW_g, and the associated deformatoin by δl_g. With g as the gravitational acceleration and ρ as the volumetric density of the object, δW_g and δl_g can

とき，δW_g と δl_g はそれぞれ式(11-10)と(11-11)で表せる．また式(11-12)のように，δl_g の積分から物体全体の自重による弾性変形量 ΔL_g が求まる．このように，自重による弾性変形量は断面形状には依存しないが，長さ L の 2 乗に比例するため，長尺の精密測定の場合には考慮を要する．100 mm の鋼製のブロックゲージ（ρ＝7850 kg/m³，E＝210 GPa）では垂直に保持しても自重による変形は 2 nm 以下とその精度に対して変形量は少ないが，1000 mm の場合は 0.18 µm となり無視できないため，100 mm 以上のブロックゲージは水平姿勢で保持することが定められている．

　一方，精密に作られた真直基準であっても，水平面に直置きして保持した場合には摩擦力による拘束が生じたり，平らな正しい水平面を用意できない場合にはその形状に応じて変形したりして形状が安定せず，基準面としての機能を発揮できない．そのため精度を要する長尺の直方体（梁）を保持する場合には対称な 2 点で支持した方が形状の再現性が高くなる．また，その変形状態を材料力学に基づく計算により求めて

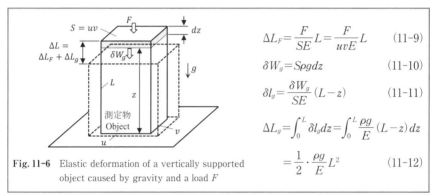

$$\Delta L_F = \frac{F}{SE}L = \frac{F}{uvE}L \qquad (11\text{-}9)$$

$$\delta W_g = S\rho g\,dz \qquad (11\text{-}10)$$

$$\delta l_g = \frac{\delta W_g}{SE}(L-z) \qquad (11\text{-}11)$$

$$\Delta L_g = \int_0^L \delta l_g dz = \int_0^L \frac{\rho g}{E}(L-z)\,dz$$

$$= \frac{1}{2}\cdot\frac{\rho g}{E}L^2 \qquad (11\text{-}12)$$

Fig. 11-6　Elastic deformation of a vertically supported object caused by gravity and a load F

be expressed by Eqs. (11-10) and (11-11), respectively. The deformation ΔL_g caused by the entire self-weight of the object can then be obtained in Eq. (11-12). Although ΔL_g is not related to the cross-section of the object, it should be considered in precision measurement of long objects as ΔL_g is proportional to L^2. ΔL_g is less than 2 nm for a vertically supported steel gauge block of L=100 mm (ρ= 7850 kg/m³, E=210 GPa). However, ΔL_g increases to 0.18 µm when L=1000 mm. For this reason, gauge blocks longer than 100 mm should be supported horizontally.

　When a precision straightedge is directly mounted on a horizontal surface, deformations of the straightedge may be caused by the friction force and the form error of the supporting surface, which will influence the function of the precision straightedge. For this reason, supporting a long straightedge (a rectangular beam) with two symmetric points is required for better form reproducibility. The locations of the supporting points can be determined by evaluating the deformation of the beam

適切な支持点を決定可能である.

Fig. 11-7 に示すように梁を対称な位置で支持する場合を考える. 梁の質量を m, 断面二次モーメントを I とする. 支持点での縦方向変形量を 0 とすると, 位置 x にある P 点の縦方向変形量 y は式(11-13), (11-14), たわみ角 θ は式(11-15), (11-16)で表される. また, 端部の変形量 y_e とたわみ角 θ_e は式(11-17), (11-18), 中央部の変形量 y_c は式(11-19)となる. これらの式に適切な条件を設定することで工学的に有用な支持点位置が求められる.

式(11-18)に $\theta_e=0$ を代入すると梁の両端面が鉛直となる支持点位置 $2a=0.5774L$ $(s=0.2113L)$ が求められる. この支持点はエアリー点と呼ばれ, 両端面間の間隔で

$$y=\frac{mg/L}{48EI}\{2x^4+3(L-4a)Lx^2-(2a^4+3(L-4a)La^2)\} \quad (0\leq x\leq a) \tag{11-13}$$

$$y=\frac{mg/L}{48EI}\{2x^4-4Lx^3+3L^2x^2-12La^2x-(2a^2-16La+3L^2)a^2\}\left(a\leq x\leq\frac{L}{2}\right) \tag{11-14}$$

$$\theta=\frac{mg/L}{24EI}\{4x^3+3(L-4a)Lx\} \quad (0\leq x\leq a) \tag{11-15}$$

$$\theta=\frac{mg/L}{24EI}(4x^3-6Lx^2+3L^2x-6La^2) \quad \left(a\leq x\leq\frac{L}{2}\right) \tag{11-16}$$

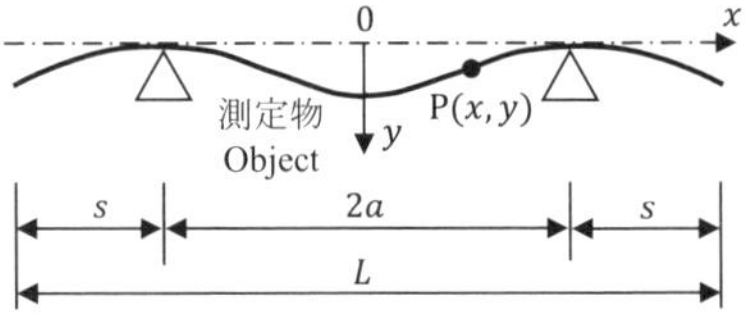

$$y_e=\frac{mg/L}{48EI}\left(\frac{3}{8}L^4-9L^2a^2+16La^3-2a^4\right) \tag{11-17}$$

$$\theta_e=\frac{mg/L}{24EI}\left(\frac{L^3}{2}-6La^2\right) \tag{11-18}$$

$$y_c=\frac{mg/L}{48EI}\{-[2a^4+3(L-4a)La^2]\} \tag{11-19}$$

Fig. 11-7 Elastic deformation of a horizontally supported object by gravity

based on the mechanics of materials.

Assume a beam shown in Fig. 11-7 has a mass m and a second moment of area I, and the vertical deformations of the beam at the supporting points are 0. The vertical deformation y and the bending angle θ of point P at position x can be expressed by Eqs. (11-13), (11-14), and Eqs. (11-15), (11-16), respectively. Those at the two ends of the beam can be expressed by Eqs. (11-17), (11-18), while the deformation y_c at the center position by Eq. (11-19). The proper supporting points can be determined based on the equations.

By substituting $\theta_e=0$ into Eq. (11-18), a result of $2a=0.5774L$ $(s=0.2113L)$ can be obtained where the beam has vertical end surfaces. Such supporting points are called Airy points, which are employed to support an end standard for defining a

長さを規定する端度器の支持点として用いられる．JIS B 7506 では，呼び寸法 100 mm
を超えるブロックゲージの寸法は，両端からそれぞれ呼び寸法の 0.211 倍の距離で適
正に支持した姿勢における寸法とすると規定されており，両端面が垂直となり寸法差
幅が小さくなるエアリー点で支持して使用する．

　$2a=0.5594L$（$s=0.2203L$）となる点は，ベッセル点と呼ばれ，中立線の両端間距
離の変化（縮み）が最小となるため，標準尺など，平行に刻まれた目盛線の間隔で長
さを規定する線度器の支持点として用いられる．

　梁の中央と両端のたわみが等しくなる支持点位置は $y_c=y_e$ の条件から $2a=0.5537L$
（$s=0.2231L$）と求められる．中央の変形量 y_c が 0 となり支持点と中央の高さが同じ
となる点は式 (11-13) に $x=0$, $y=0$ を代入して解くことで $2a=0.5227L$（$s=0.2386L$）
と求められる．

　接触式プローブで試料表面を測定する場合，測定力が試料表面の小さい面積に集中
して作用するので変形量が顕著になり，試料表面にダメージを生じさせる可能性があ
る．特に触針式表面粗さ測定機の場合，ダイヤモンド触針の先端半径が μm オーダと
小さいため，触針先端による試料表面のミクロな変形に特に注意を要する．ヘルツの

length by the distance between the two end surfaces. In JIS B 7506 (ISO 3650), the
length of a gauge block with a nominal length L longer than 100 mm is defined as
the length of the gauge block when it is supported at two points at a distance of
$0.211L$ from each end (the Airy points).

The supporting points under the condition of $2a=0.5594L$ ($s=0.2203L$) are
called the Bessel points where the length of the beam center line has the minimum
change. Line scales, which define lengths by the intervals between parallel graduations, are supported at such points.

The supporting points under the condition of $2a=0.5537L$ ($s=0.2231L$) can be
obtained by taking $y_c=y_e$ for the same deformations at the center and ends of the
beam. The supporting points when $2a=0.5227L$ ($s=0.2386L$) can be obtained by
substituting $x=0$, $y=0$ into Eq. (11-13) for $y_c=0$ and the same vertical positions at
the supporting points and the center point.

When the surface of an object is detected by a contact probe, the deformation
may become significant to damage the object surface since the measuring force is
concentrated in a small area of the surface. This is especially the case in surface
roughness measurement using a diamond stylus with a micrometric tip radius. Such
a deformation can be analyzed with Hertz's elastic contact theory. As shown in

弾性接触理論を用いることによって，このような変形を解析することができる．Fig.11-8に示すように，測定対象は平面（曲率半径：∞），触針の先端半径を r_{tip}，測定力 F をとると，接触円半径 q，および試料表面の弾性変形量 δ が式(11-20)から式(11-22)のように求められる．式(11-22)の E_1, ν_1, E_2, ν_2 はそれぞれ触針と試料のヤング率とポアソン比で，E^* は等価ヤング率である．また，触針による平均接触圧 P_{mean} は式(11-23)で表され，これが試料の降伏強さ（アルミ合金など非鉄材料の場合は0.2%耐力）の3倍より大きくなると試料表面に塑性変形が起こる．

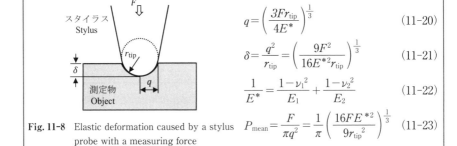

$$q=\left(\frac{3Fr_{tip}}{4E^*}\right)^{\frac{1}{3}} \quad (11\text{-}20)$$

$$\delta=\frac{q^2}{r_{tip}}=\left(\frac{9F^2}{16E^{*2}r_{tip}}\right)^{\frac{1}{3}} \quad (11\text{-}21)$$

$$\frac{1}{E^*}=\frac{1-\nu_1^2}{E_1}+\frac{1-\nu_2^2}{E_2} \quad (11\text{-}22)$$

$$P_{mean}=\frac{F}{\pi q^2}=\frac{1}{\pi}\left(\frac{16FE^{*2}}{9r_{tip}^2}\right)^{\frac{1}{3}} \quad (11\text{-}23)$$

Fig. 11-8　Elastic deformation caused by a stylus probe with a measuring force

Fig. 11-8, denoting the tip radius of the stylus by r_{tip}, the measuring force by F, the contacting circle radius q and the elastic deformation δ can be expressed by Eqs. (11-20)-(11-22). E_1, ν_1, E_2, ν_2 in Eq. (11-22) are the Young's moduli and the Poisson ratios of the stylus and a plane object surface, respectively. E^* is the effective Young's modulus. The mean pressure of the stylus against the object surface is expressed by P_{mean} in Eq. (11-23). The object surface will be damaged if P_{mean} is larger than three times the yield strength (or the 0.2% offset yield strength for a nonferrous material such as an aluminum alloy) of the object.

【演習問題】

11-1) スケールと物体のミスアライメント角度を 1° として，1000 mm の物体を測った場合のコサイン誤差 δ_{cos} 求めよ．

11-2) スケールと物体のアッベオフセットを 100 mm とし，視線角度誤差を 1° として，アッベ誤差 δ_{Abbe} 求めよ．

11-3) 23℃ の環境において，ステンレス鋼製スケールで 1000 mm 長さのインバー製直方体を測ったときの読み R' を示せ（Fig. 11-5）．

11-4) 長さ 1500 mm 直方体を支持するエアリー点の位置を計算せよ（Fig. 11-7）．

11-5) 測定力 0.1 mN，先端半径 5 μm のダイヤモンド触針がアルミ合金 A2017 の平板に接触した際の弾性変形量 δ を計算せよ．また，塑性変形が生じるかどうかを判定せよ．

【Problems】

11-1) Assume the misalignment angle between a scale and an object is 1°. Calculate the cosine error δ_{cos} when the length of the object is 1000 mm.

11-2) Assume the Abbe offset between a scale and an object is 100 mm. Calculate the cosine error δ_{Abbe} if the deviation angle of the light of sight is 1°.

11-3) Under a temperature of 23℃, show the reading R' of a stainless steel scale used to measure a rectangular object made of invar with a length of 1000 mm (Fig. 11-5).

11-4) Obtain the positions of the Airy supporting points in Fig. 11-7 for a rectangular object 1500 mm long.

11-5) An aluminum alloy A2017 surface is detected by a diamond stylus with a tip radius of μm. Calculate the elastic deformation δ and judge if the aluminum alloy surface will be damaged when the measuring force is 0.1 mN.

第12章　自律校正法

精密計測においては，事前に上位測定標準によって校正され，測定の不確かさが十分に小さい基準との比較によって信頼性の高い測定値が得られる．例えば，直動ステージの真直度誤差測定の場合，上位標準の光波干渉計などで形状偏差が校正された高精度直定規が計測の基準となる．一方，超精密直動ステージの計測において，求められる真直度の計測不確かさは直定規の校正結果の不確かさと同程度，あるいはより小さくなる場合がある．また，大型工作機械の長ストローク直動軸を計測する場合，十分に長い直定規が入手できなくなるため，形状誤差が未知の加工物を直定規の代わりに用いる必要が生じる．これらの課題に対応するため，測定標準など高精度な外部基準を必要としない自律校正が有効となる．本章では，誤差分離法による形状と運動誤差の自律校正について述べる．

Chapter 12　Self-Calibration Methods

In precision metrology, a reliable measurement result is obtained by comparison against an accurate reference with negligible measurement uncertainty, which is precalibrated with respect to a higher-level measurement standard. For instance, in the measurement of the straightness motion error of a linear stage, a high-precision straightedge, of which form error is precalibrated by an optical surface interferometer, is employed as the reference. However, in the measurement of an ultraprecision linear stage, the required measurement uncertainty of motion error can be comparable to or smaller than that of the form error. In the measurement of the long-stroke linear axis of a large-scale machine tool, utilizing a machined workpiece with an unknown form error for the measurement is recommended, as long straightedges are unavailable. In such cases, self-calibration without using an accurate reference is an effective solution. In this chapter, the self-calibration of surface profiles and stage motion errors based on the error separation methods is presented.

● 12.1　マルチステップ法

　誤差分離法による自律校正では，測定対象と誤差を含む連立方程式を立て，その方程式を数学的に解くことによって測定対象を誤差から分離して求める．連立方程式の立て方の1つとして，マルチステップ法がある．この方法では，測定条件を変更した計測を複数回行うことで，必要な数の式を立てるようにしている．

　マルチステップ法の1つとして，フィゾー干渉計の参照平面（オプティカルフラット）を自律校正する3枚合わせ法がある．Fig. 12-1 に示すように，校正対象のオプティカルフラット A と同一寸法のオプティカルフラット B，C を用意する．それらの断面形状をそれぞれ $a(x)$，$b(x)$，$c(x)$とする．ステップ I では，A を参照面にして B を測定し，$a(x)$と $b(x)$の和からなる干渉計の出力 $m_1(x)$を得る．ステップ II と III では，それぞれ C と B，C と A の組み合わせで測定を行い，干渉計の出力 $m_2(x)$と $m_3(x)$

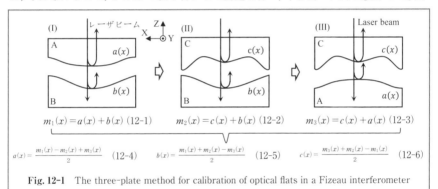

Fig. 12-1　The three-plate method for calibration of optical flats in a Fizeau interferometer

● 12.1　Multi-step methods

　For self-calibration based on error-separation, a set of simultaneous equations are first established. Each equation comprises the measurement and reference quantities, which can be obtained by solving the equations. The multi-step method is one of the ways to establish simultaneous equations where measurements are carried out under different conditions.

　As one of the multi-step methods, the three-plate method is employed for self-calibration of the reference plate (optical flat) of the Fizeau interferometer. As shown in Fig. 12-1, in addition to reference plate A, two identical optical flats B and C are employed. The sectional surface profile errors of the plates are denoted by $a(x)$, $b(x)$, and $c(x)$. In Step I, B is measured with reference plate A. An interferometer output $m_1(x)$ is obtained, which is the sum of $a(x)$ and $b(x)$. In Steps II and III, interferometer outputs $m_2(x)$ and $m_3(x)$ are obtained with the combination of C and B, and that of C and A, respectively. As shown in Eqs. (12-4)-(12-6), $a(x)$,

を得る. 式(12-1)〜(12-3)を連立して解くことによって, 式(12-4)〜(12-6)のように, 断面形状誤差 $a(x)$, $b(x)$, $c(x)$ を干渉計の出力から分離して求めることができる. また, ステップ III において, A を光軸を中心に回転することによって, A の異なる断面の形状を求めることが可能となる. この計測法ではオプティカルフラットの変形やアライメント誤差などが主な測定不確かさ要因となる.

Fig. 12-2 に示すマルチステップ法によって, XY 平面回折格子の Z 方向平面度 $e_Z(x, y)$ のみならず, X と Y 方向の格子ピッチ偏差 $e_X(x, y)$, $e_Y(x, y)$ も複数のステップで得られるフィゾー干渉計の出力から分離して自律校正することができる. 図には平面格子からの 0 次と X 方向±1 次回折光が示されており, θ は式(12-7)で定義される回折角である. ステップ I では, 格子面は干渉計の参照面と平行になるように設置

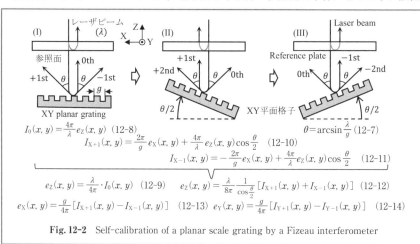

$$I_0(x, y) = \frac{4\pi}{\lambda} e_Z(x, y) \quad (12\text{-}8)$$

$$\theta = \arcsin \frac{\lambda}{g} \quad (12\text{-}7)$$

$$I_{X+1}(x, y) = \frac{2\pi}{g} e_X(x, y) + \frac{4\pi}{\lambda} e_Z(x, y) \cos \frac{\theta}{2} \quad (12\text{-}10)$$

$$I_{X-1}(x, y) = -\frac{2\pi}{g} e_X(x, y) + \frac{4\pi}{\lambda} e_Z(x, y) \cos \frac{\theta}{2} \quad (12\text{-}11)$$

$$e_Z(x, y) = \frac{\lambda}{4\pi} \cdot I_0(x, y) \quad (12\text{-}9)$$

$$e_Z(x, y) = \frac{\lambda}{8\pi} \frac{1}{\cos^2 \frac{\theta}{2}} [I_{X+1}(x, y) + I_{X-1}(x, y)] \quad (12\text{-}12)$$

$$e_X(x, y) = \frac{g}{4\pi} [I_{X+1}(x, y) - I_{X-1}(x, y)] \quad (12\text{-}13)$$

$$e_Y(x, y) = \frac{g}{4\pi} [I_{Y+1}(x, y) - I_{Y-1}(x, y)] \quad (12\text{-}14)$$

Fig. 12-2 Self-calibration of a planar scale grating by a Fizeau interferometer

$b(x)$, and $c(x)$ can be obtained from the outputs in Eqs. (12-1)-(12-3). In Step III, the profile errors in different sections of the surface of A can be obtained by rotating A about the optical axis. The deformation and alignment errors of the optical flats are the major uncertainty sources.

A Fizeau interferometer is typically employed for out-of-flatness measurements. With the multi-step method shown in Fig. 12-2 for self-calibration of an XY planar diffraction grating, not only the Z-directional out-of-flatness $e_Z(x, y)$ but also the X- and Y-directional pitch deviations $e_X(x, y)$, $e_Y(x, y)$ can be measured using the outputs of the Fizeau interferometer. The 0th and the ±1st order diffraction beams from the grating are illustrated in the figure where θ is the diffraction angle defined by Eq. (12-7). In Step I, the grating is aligned to be parallel to the reference surface of the interferometer for obtaining the intensity $I_0(x, y)$ of the interference

し，格子面からの 0 次光と参照面からの参照光の干渉強度 $I_0(x, y)$ を得る．参照面の形状誤差が格子の平面度誤差 $e_Z(x, y)$ に比べて無視できるほど小さいとき，$I_0(x, y)$ は式(12-8)のように表され，$e_Z(x, y)$ は式(12-9)のように直接 $I_0(x, y)$ から求めることができる．一方，ステップ II と III では，格子面を図に示すように $\theta/2$ だけ傾け，±1 次光と参照光の干渉強度 $I_{X+1}(x, y)$ と $I_{X-1}(x, y)$ をそれぞれの式(12-10)，(12-11)のように得る．$e_Z(x, y)$ は同位相の成分，また $e_X(x, y)$ は逆位相の成分を $I_{X+1}(x, y)$ と $I_{X-1}(x, y)$ にそれぞれ生じさせるため，式(12-12)，(12-13)のように $e_Z(x, y)$ と $e_X(x, y)$ を分離することができる．

同様に Y 方向±1 次回折光と参照光との干渉強度 $I_{Y+1}(x, y)$，$I_{Y-1}(x, y)$ を利用すれば，式(12-14)のように，Y 方向ピッチ偏差 $e_Y(x, y)$ を求められる．また，参照面の形状誤差 $a(x, y)$ が無視できない場合は，式(12-8)，(12-10)，(12-11)の干渉強度には $a(x, y)$ が含まれることとなる．演算がやや煩雑となるが，この 3 つの式から，$a(x, y)$，$e_Z(x, y)$，$e_X(x, y)$ を分離できる．回折格子 1 枚でフィゾー干渉計の参照面が自律校正できるので，この方法は前述の 3 枚合わせ法に比べてより簡便に実施できる利点がある．この方法の格子の設置誤差などが主な測定不確かさ要因となる．

signal between the 0th-order beam from the grating and the reference beam from the reference surface. Assuming that the out-of-flatness of the reference surface is sufficiently small compared with $e_Z(x, y)$, $I_0(x, y)$ can be expressed by Eq. (12-8), from which $e_Z(x, y)$ can be directly obtained in Eq. (12-9). In Steps II and III, the grating is tilted $\theta/2$ as shown in the figures for obtaining the interference intensities $I_{X+1}(x, y)$ (Eq. (12-10)) and $I_{X-1}(x, y)$ (Eq. (12-11)) between the ±1st-order beams and the reference beam. $e_Z(x, y)$ and $e_X(x, y)$, which introduce in-phase and out-of-phase components in Eqs. (12-10) and (12-11), respectively, can then be obtained in Eqs. (12-12) and (12-13).

The intensities $I_{Y+1}(x, y)$, $I_{Y-1}(x, y)$ of the interference signals between the Y-directional 1st-order beams and the reference beam, which are not shown in Fig. 12-2, can be employed to evaluate the Y-directional pitch deviation $e_Y(x, y)$ in the same way by Eq. (12-14). Meanwhile, if the out-of-flatness $a(x, y)$ of the reference surface is not sufficiently small, $a(x, y)$ will introduce the corresponding components in Eqs. (12-8), (12-10) and (12-11). Although the procedure is more complicated, $a(x, y)$, $e_Z(x, y)$, $e_X(x, y)$ can be mathematically evaluated from the three equations. Since only one XY grating is needed for calibration of the reference surface of a Fizeau interferometer, this method is simpler than the three-plate method. The alignment error of the grating is the major measurement uncertainty source.

　反転法は走査型形状測定における代表的なマルチステップ法の1つである．Fig. 12-3に反転法による真円度の測定原理を示す．真円度測定の場合，まずステップIにおいて円形試料をスピンドルで1周回転させて変位プローブの出力$m_1(\theta)$を得る．式(12-15)のように，$m_1(\theta)$は試料の真円度誤差$r(\theta)$とスピンドルの半径方向回転誤差$s(\theta)$の和となる．ステップIIでは，試料とプローブをスピンドルの開始位置に対して180°反転させた後，再度試料を回転させてプローブ出力$m_2(\theta)$を得る．反転走査によって，式(12-16)のように$m_2(\theta)$における$s(\theta)$の位相が反転するため，$r(\theta)$と$s(\theta)$は式(12-17)，(12-18)のように求められる．Fig. 12-4のように，同様な考え方で真直度形状偏差$f(x)$を直動ステージの運動誤差$s(x)$から分離できる（式(12-19)～(12-22)）．

　また，Fig. 12-5に示す転写反転法を用いて，ダイヤモンド工具やインデンターなど

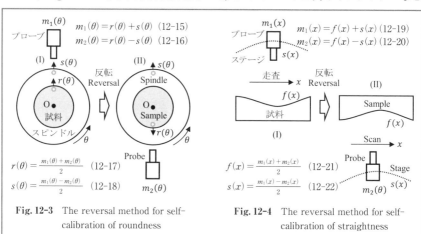

Fig. 12-3　The reversal method for self-calibration of roundness

Fig. 12-4　The reversal method for self-calibration of straightness

$$m_1(\theta) = r(\theta) + s(\theta) \quad (12\text{-}15)$$
$$m_2(\theta) = r(\theta) - s(\theta) \quad (12\text{-}16)$$
$$r(\theta) = \frac{m_1(\theta) + m_2(\theta)}{2} \quad (12\text{-}17)$$
$$s(\theta) = \frac{m_1(\theta) - m_2(\theta)}{2} \quad (12\text{-}18)$$

$$m_1(x) = f(x) + s(x) \quad (12\text{-}19)$$
$$m_2(x) = f(x) - s(x) \quad (12\text{-}20)$$
$$f(x) = \frac{m_1(x) + m_2(x)}{2} \quad (12\text{-}21)$$
$$s(x) = \frac{m_1(x) - m_2(x)}{2} \quad (12\text{-}22)$$

　The reversal method is one of the typical multi-step methods for probe-scan systems. The principle of the reversal method for the roundness measurement is shown in Fig. 12-3. In Step I for roundness measurement, the cylindrical sample with a roundness error $r(\theta)$ is rotated by a spindle with a spindle radial error $s(\theta)$ for obtaining the output $m_1(\theta)$ of a displacement probe over one rotation, which is the sum of $r(\theta)$ and $s(\theta)$ (Eq. (12-15)). In Step II, after the sample and the probe are reversed 180° at the starting position of the spindle, the sample is rotated to acquire probe output $m_2(\theta)$ (Eq. (12-16)) with $s(\theta)$ of out-of-phase. $r(\theta)$ and $s(\theta)$ can then be obtained in Eqs. (12-17) and (12-18). Similarly, the straightness surface profile error $f(x)$ can be separated from the straightness motion error of a linear stage by the reversal method shown in Fig. 12-4 (Eqs. (12-19)-(12-22)).

　The tip radius R_s of a sample made of a hard material, such as a diamond cutting tool or a diamond stylus, can be accurately measured using the replication reversal

硬い材質のサンプルの先端半径 R_s を高精度に計測することができる．この方法では，サンプルの先端形状を軟質金属に転写して同じ先端半径 R_s を持つ反転レプリカを作り，先端半径 R_p のプローブでサンプルとレプリカをそれぞれ走査して式(12-23)と(12-24)に示す半径出力 R_{m1} と R_{m2} を得る．その結果，式(12-25)と(12-26)のように R_s と R_p を分離して求めることができる．なお，この方法はサンプルおよびプローブ先端の形状誤差の分離にも適用できる．

反転前後の運動誤差の再現性が反転法の主な測定不確かさ要因になる．それを低減させるためのマルチ方位法（Fig. 12-6）では，スピンドルの開始位置に対する試料の相対方位角を，1周を N 等分した角度 $\Delta\phi$（式(12-27)）で分割する．$i\Delta\phi(i=0,1,\cdots,N-1)$ の方位に対応するプローブ出力 $m_i(\theta)$ を式(12-28)のように取得する．分割数

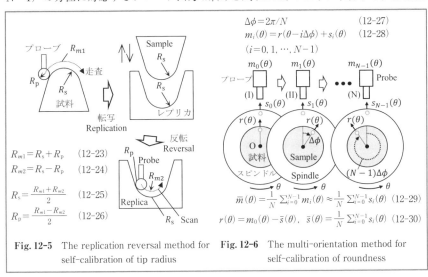

Fig. 12-5 The replication reversal method for self-calibration of tip radius

Fig. 12-6 The multi-orientation method for self-calibration of roundness

method shown in Fig. 12-5. In this method, a replica of the sample tip, which has the same radius R_s, is made onto a soft metal by indentation. The sample and the replica are scanned by the probe with a tip radius of R_p, from which the radii R_{m1} and R_{m2} shown in Eqs. (12-23) and (12-24) can be obtained. Consequently, R_s and R_p can be evaluated from Eqs. (12-25) and (12-26), respectively. This method can also be applied to separate the tip profiles of the sample from the probe.

The multi-orientation method of roundness measurement shown in Fig. 12-6 can be employed to reduce the measurement uncertainty caused by the nonrepeatable components of the spindle error in the reversal method. The orientations of the sample over 360° are equally divided by N for an equal interval $\Delta\phi$ (Eq. (12-27)). The probe output $m_i(\theta)$ for the sample set at $i\Delta\phi(i=0, 1, \cdots, N-1)$ is expressed in

N の整数倍の調和次数成分を無視すれば，式(12-29)が得られる．式(12-30)では $s(\theta)$ の平均値から $r(\theta)$ を計算しているので，$s(\theta)$ の再現性の影響が低減される．一方，この方法では分割数 N の整数倍の調和次数成分が測定できないという問題があり，測定不確かさの要因となる．

Fig.12-7 に回転走査が困難な小径円筒サンプルを測定するためのマルチ方位法を示す．スティッチング直線走査法と呼ばれるこの方法では，サンプルの設置方位角を $\alpha(=2\pi/N)$ ずつ変更させた状態で x 方向に直線走査をし，それによって N 個の円弧形状 $f_i(x)\,(i=1,2,\cdots,N)$ を得る．$f_i(x)$ から i 番目円弧の平均半径 $\bar{R}_i$ を計算し，それをもとに N 個の円弧の平均半径 $\bar{R}_a\!\left(=\dfrac{1}{N}\displaystyle\sum_{i=1}^{N}\bar{R}_i\right)$ を算出する．極座標系に変換された円

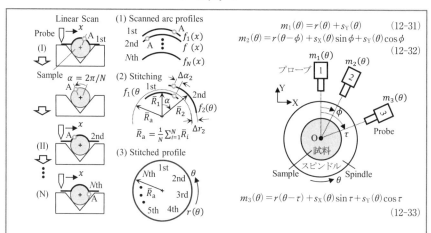

Fig. 12-7　The stitching linear-scan method for accurate measurement of radius and roundness　**Fig. 12-8**　The three-probe method for self-calibration of roundness

$$m_1(\theta) = r(\theta) + s_Y(\theta) \qquad (12\text{-}31)$$
$$m_2(\theta) = r(\theta-\phi) + s_X(\theta)\sin\phi + s_Y(\theta)\cos\phi \qquad (12\text{-}32)$$
$$m_3(\theta) = r(\theta-\tau) + s_X(\theta)\sin\tau + s_Y(\theta)\cos\tau \qquad (12\text{-}33)$$

Eq. (12-28). By ignoring the harmonics of integer multiples of N, Eq. (12-29) can be obtained where the nonrepeatable components of $s(\theta)$ are reduced, followed by the result of $r(\theta)$ in Eq. (12-30). The harmonics of integer multiples of N cannot be measured, which is one of the major measurement certainties of this method.

Fig. 12-7 shows a multi-orientation method, called the stitching linear-scan method, for a cylindrical sample with a small diameter. Due to the challenges of conducting rotary scans on such a sample, this method involves linear scanning of the sample along the X-axis. The orientation of the sample is changed by an angle of $\alpha(=2\pi/N)$ before each linear scan for obtaining N sets of arc profiles $f_i(x)\,(i=1, 2, \cdots, N)$. The mean radius $\bar{R}_a\!\left(=\dfrac{1}{N}\displaystyle\sum_{i=1}^{N}\bar{R}_i\right)$ of the arc profiles can be obtained, where $\bar{R}_i$ is the mean radius of $f_i(x)$. The offset errors Δr_i of sample alignment can be removed by taking $\bar{R}_a$ as a reference in the stitching of the arc profiles $f_i(\theta)$ in the polar coor-

弧形状 $f_i(\theta)$ $(i=1, 2, \cdots, N)$ に対して，$\bar{R}_a$ を基準に半径方向のスティッチングを施すことによって，サンプル設置のオフセット誤差 Δr_i を自律的に除去する．また隣接の円弧形状が一致する部分を基準に円周方向のスティッチングを施すことによって，方位角設置誤差 $\Delta\alpha_i$ を自律的に補正した形でサンプルの平均半径 $\bar{R}_a$ と真円度形状 $r(\theta)$ を高精度に求めることができる．

● 12.2 マルチプローブ法

Fig. 12-8 に示す 3 点法のように，マルチプローブ法では 1 回の走査で得られる複数のプローブ出力を用いて誤差分離に必要な連立方程式を立てる．真円度測定の場合，真円度誤差 $r(\theta)$ と回転誤差の XY 方向成分 $s_X(\theta)$，$s_Y(\theta)$ という 3 つの未知数がプローブの出力に含まれる（式(12-31)～(12-33)）．Fig. 12-9 のように，まず変位プローブの出力 $m_1(\theta)$, $m_2(\theta)$, $m_3(\theta)$ を用いて $r(\theta)$ のみの関数となる差動出力 $\Delta m_3(\theta)$（式(12-34)）を算出する．$\Delta m_3(\theta)$ と $r(\theta)$ のフーリエ変換をそれぞれ $\Delta M_3(n)$ と $R(n)$ とすると，式(12-35)の関係式が得られる．ここで n は空間周波数を表す調和次数であり，1 周における波の数を示している．$\Delta M_3(n)$ と $R(n)$ の商となる 3 点法の伝達関数 $H_3(n)$ は式(12-36)になるので，$R(n)$ は $\Delta M_3(n)$ と $H_3(n)$ から求められ，$r(\theta)$ は $R(n)$ の逆フーリ

dinate system along the radial direction. Similarly, the orientation errors $\Delta\alpha_i$ of sample alignment can be removed in the stitching along the circumferential direction since the overlapped areas of the arc profiles are the same. Consequently, the mean radius $\bar{R}_a$ and the roundness $r(\theta)$ can be accurately obtained.

● 12.2 Multi-probe methods

As in the three-probe method of roundness measurement shown in Fig. 12-8, simultaneous equations for error-separation can be established using the outputs of multiple probes acquired in a single scan. In roundness measurement, three unknowns are included in each probe output: the roundness error $r(\theta)$, and the X- and Y-directional spindle error components $s_X(\theta)$, $s_Y(\theta)$ (Eq. (12-31)-(12-33)). As shown in Fig. 12-9, the displacement probe outputs $m_1(\theta)$, $m_2(\theta)$, and $m_3(\theta)$ are employed to obtain a differential output $\Delta m_3(\theta)$ (Eq. (12-34)) in which $s_X(\theta)$ and $s_Y(\theta)$ are removed and only $r(\theta)$ is included. Assuming the Fourier transforms of $\Delta m_3(\theta)$ and $r(\theta)$ are denoted by $\Delta M_3(n)$ and $R(n)$, Eq. (12-35) can be obtained where n is the harmonic number (the spatial frequency showing the number of undulations per revolution). The transfer function $H_3(n)$ of the three-probe method, which is the quotient of $\Delta M_3(n)$ and $R(n)$, is shown in Eq. (12-36). $R(n)$ can then be

工変換から算出できる．さらに，$r(\theta)$をプローブの出力に代入することで，$s_X(\theta)$と$s_Y$$(\theta)$が計算できる．

　伝達関数$H_3(n)$は３点法における各調和次数の感度を表している．$H_3(n)$の振幅が0になる，つまり感度が0になる次数が複数存在し，これらの次数における真円度および回転誤差の周波数成分が計測できない．これは変位4点法や角度3点法など，変位プローブのみあるいは角度プローブのみを用いるすべてのマルチプローブ法に存在する基本的な問題である．最大で3本の変位プローブと角度プローブを用いる混合法によってこの問題を根本的に解決できる．また，Fig. 12-10に示す直交型混合法では，直角に配置した変位プローブと角度プローブの出力 $m_1(\theta)$, $\mu_2(\theta)$（式(12-37)）から，最

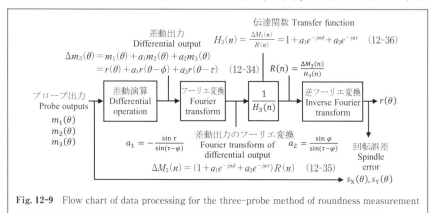

Fig. 12-9　Flow chart of data processing for the three-probe method of roundness measurement

obtained from the quotient of $\Delta M_3(n)$ and $H_3(n)$, and $r(\theta)$ can be obtained from the inverse Fourier transform of $R(n)$. $s_X(\theta)$ and $s_Y(\theta)$ can also be evaluated by substituting $r(\theta)$ into the probe outputs.

　The transfer function $H_3(n)$ represents the harmonic sensitivity of the three-probe method. Frequencies exist at which the amplitude of $H_3(n)$, i.e., the harmonic sensitivity, becomes zero, where the harmonic components of the roundness error and the spindle error will be lost. This is called the harmonic-loss problem, which is a fundamental problem existing in all kinds of multi-probe methods using only displacement probes or slope probes, including the four-displacement probe method and the three-slope probe method. This problem can be completely solved by the mixed method, which uses maximally three displacement and slope probes. The orthogonal mixed method uses the outputs $m_1(\theta)$ and $\mu_2(\theta)$ (Eq. (12-37)) of a displacement probe and a slope probe for obtaining the differential output $m_{om}(\theta)$ (Eq. (12-38)) in a simplest way as shown in Fig. 12-10. The transfer function $H_{om}(n)$ of this method (Eq. (12-39)) without any harmonics of zero sensitivity is shown in

も簡単な形で差動出力 $m_{\mathrm{om}}(\theta)$ を得る（式(12-38)）．Fig.12-11 に感度が0になる次数が存在しない直交型混合法の伝達関数 $H_{\mathrm{om}}(n)$（式(12-39)）を示す．プローブ間の感度違いとプローブ設置誤差がマルチプローブ法の主な不確かさ要因となる．

●12.3　ハイブリッド法

　マルチプローブ法とマルチステップ法，あるいはそれらの方法と外部センサとの組み合わせを利用するハイブリッド法がある．Fig.12-12に2点法と外部オートコリメータをハイブリッドした真直度測定を示す．外部オートコリメータによって走査ステージ Z 軸周りのチルト誤差 $s_t(x)$ を直接計測し補正する（式(12-42)）．さらにプローブ間隔が d の変位プローブ A，B の出力（式(12-40)，(12-41)）を用いて Z 軸方向の並進誤差 $s_Z(x)$ を除去した差動出力 $\Delta m_2(x)$（式(12-43)）を計算し，その一階積分から式

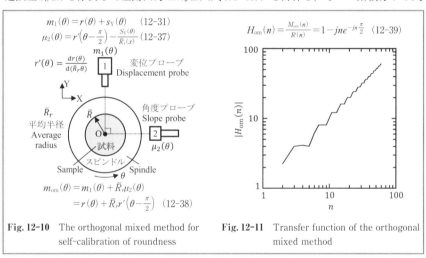

Fig. 12-10　The orthogonal mixed method for self-calibration of roundness

Fig. 12-11　Transfer function of the orthogonal mixed method

Fig. 12-11. The mismatching of probe sensitivities and the alignment errors of probes are the major measurement uncertainties of multi-probe methods.

●12.3　Hybrid methods

Combination of a multi-probe method with a multi-step method, or with an external sensor can be employed as a hybrid method for error-separation. Fig. 12-12 shows a hybrid system that combines the two-probe method and an external autocollimator for straightness measurement. The tilt error motion $s_t(x)$ of the scanning stage about the Z-axis is detected by the autocollimator for compensation (Eq. (12-42)). The translational error motion $s_Z(x)$ about the Z-axis is removed by taking the differential output $\Delta m_2(x)$ of the two probes A and B with a separation d as shown in

(12-44) のように試料の形状 $f(x)$ が復元される.

一方,Fig. 12-13 のように,3 点法を真直度測定にそのまま展開したとき,それぞれのプローブ出力(式(12-45)~(12-47))のゼロ点調整時に存在する誤差 e_1, e_2, e_3 によって,式(12-48)の 3 点法の差動出力にゼロ点差 $\alpha(=e_3-2e_2+e_1)$ の項を生じさせてしまう.式(12-49)のように,$\Delta m_3(x)$ の 2 回積分演算から試料の形状 $f(x)$ を復元する際,α に比例する放物線状誤差成分が生じる問題が発生する.この放物線状誤差成分は試料の測定長の 2 乗に比例するので,3 点法による長尺試料測定における最大の不

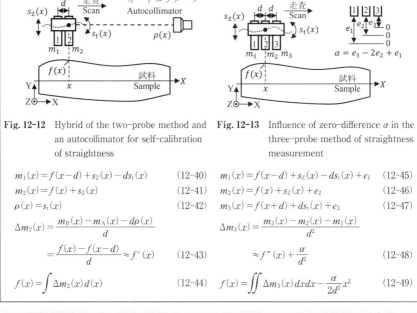

Fig. 12-12 Hybrid of the two-probe method and an autocollimator for self-calibration of straightness

Fig. 12-13 Influence of zero-difference α in the three-probe method of straightness measurement

$$m_1(x) = f(x-d) + s_Z(x) - ds_t(x) \qquad (12\text{-}40)$$

$$m_2(x) = f(x) + s_Z(x) \qquad (12\text{-}41)$$

$$\rho(x) = s_t(x) \qquad (12\text{-}42)$$

$$\Delta m_2(x) = \frac{m_B(x) - m_A(x) - d\rho(x)}{d}$$

$$= \frac{f(x) - f(x-d)}{d} \approx f'(x) \qquad (12\text{-}43)$$

$$f(x) = \int \Delta m_2(x)\, d(x) \qquad (12\text{-}44)$$

$$m_1(x) = f(x-d) + s_Z(x) - ds_t(x) + e_1 \qquad (12\text{-}45)$$

$$m_2(x) = f(x) + s_Z(x) + e_2 \qquad (12\text{-}46)$$

$$m_3(x) = f(x+d) + ds_t(x) + e_3 \qquad (12\text{-}47)$$

$$\Delta m_3(x) = \frac{m_3(x) - m_2(x) - m_1(x)}{d^2}$$

$$\approx f''(x) + \frac{\alpha}{d^2} \qquad (12\text{-}48)$$

$$f(x) = \iint \Delta m_3(x)\, dx dx - \frac{\alpha}{2d^2} x^2 \qquad (12\text{-}49)$$

Eqs. (12-40), (12-41) and from the integral of which the surface profile $f(x)$ can be obtained.

On the other hand, as shown in Fig. 12-13, when the three-probe method is applied to straightness measurement, zero-adjustment errors e_1, e_2, and e_3 of the probes will introduce a term of zero-difference $\alpha(=e_3-2e_2+e_1)$ in the differential output $\Delta m_3(x)$ of the three-probe method, as shown in Eqs. (12-45)-(12-48). As can be seen in Eq. (12-49), a parabolic error term proportional to α will be generated when the surface profile $f(x)$ of the sample is calculated from the double integration of $\Delta m_3(x)$. Since this parabolic error term is proportional to the measurement length of the sample, it is the maximum uncertainty source in the straightness measurement of long samples. Fig. 12-14 shows the hybrid of the re-

確かさ要因となる.

Fig. 12-14 に反転法を組み合わせることによって，特別高精度な平面基準や補助試料を用いずにゼロ点差 α を自律的に求めるハイブリッドシステムを示す.図のように，もう1組の3点法プローブユニットを同じステージに搭載し，試料の両側の表面形状 $f(x_i)$ と $h(x_i)$ を同時に走査して測定する.X軸周りに反転される前後の試料をそれぞれ走査して得られたプローブの出力から式(12-50)と(12-51)が得られるため，それらに基づいて2つのプローブユニットのゼロ点差 α と β が計算される.これらの式における平均演算で偶然誤差による測定不確かさが低減できる.さらに式(12-49)に α を代入し補正する.図の s はサンプリング間隔で，L は測定長である.

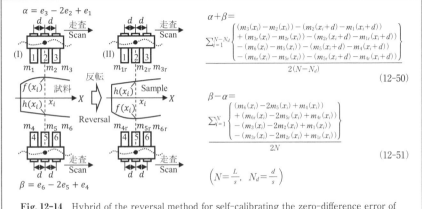

Fig. 12-14 Hybrid of the reversal method for self-calibrating the zero-difference error of three-probe method

versal method for self-calibration of the zero-difference α associated with the three-probe method instead of using any external flat reference surfaces or auxiliary artifacts. Another three-probe unit with the same probe separation d is added. The pair of probe units are moved by the same stage to simultaneously scan the surface profiles $f(x_i)$ and $h(x_i)$ of the two sides of the sample. Eqs. (12-50) and (12-51) can be established using the probe outputs acquired in the two scans before and after the reversal of the sample about the X-axis, where s is the sampling interval and L is the entire measurement length. The averaging operations in the equations reduce the measurement uncertainties caused by random errors. The zero-difference errors α and β of the two probe units can then be calculated from the two equations for compensation of the parabolic error term in Eq. (12-49).

【演習問題】

12-1) Fig. 12-2 において，参照平面の形状誤差を $a(x, y)$ とした場合，$I_0(x, y)$，$I_{X+1}(x, y)$，$I_{X-1}(x, y)$ の式を書き直せ．ただし，回折格子の傾きによる計測点の x 座標の変化も考えよ．

12-2) Fig. 12-3 の反転法において，2 回の走査における回転誤差の再現性から受ける影響を式で示せ．

12-3) Fig. 12-6 のマルチ方位法において，分割数 N の整数倍の調和次数成分が測定できないことを式で示せ．

12-4) Fig. 12-8 の真円度 3 点法において，$\phi = 10°$，$\tau = 30°$ の場合と，$\phi = 10°$，$\tau = 34°$ の場合とで，感度が 0 になる最初の調和次数はどちらが高いかを説明せよ．

12-5) Fig. 12-13 の真直度 3 点法において，$a = 10\,\text{nm}$ としたときの放物線状誤差の大きさを示せ．ただし，$L = 600\,\text{mm}$，$d = 10\,\text{mm}$ とする．

【Problems】

12-1) Assuming the form error of the reference surface in Fig. 12-2 is $a(x, y)$, rewrite $I_0(x, y)$, $I_{X+1}(x, y)$ and $I_{X-1}(x, y)$. The changes in the x coordinates of the measurement points caused by the tilt angle of the diffraction grating should also be considered.

12-2) For the reversal method shown in Fig. 12-3, investigate the influence of the nonrepeatability of the spindle error by equations.

12-3) For the multi-orientation method shown in Fig. 12-6, use equations to demonstrate that the harmonics of integer multiples of N cannot be measured.

12-4) For the three-probe method shown in Fig. 12-8, show which case has a higher first harmonic with zero sensitivity, $\phi = 10°$, $\tau = 34°$ or $\phi = 10°$, $\tau = 30°$.

12-5) For the three-probe method shown in Fig. 12-13, show the magnitude of the parabolic error when $a = 10\,\text{nm}$, $L = 600\,\text{mm}$, and $d = 10\,\text{mm}$.

第13章 機械学習と精密計測

　機械学習が一般的に用いられるようになり様々な新しい取り組みがなされている．精密計測の分野においても，機械学習が導入されており新しい進展が見られた．一方，精密計測における測定結果の裏付けとして GUM で定義されている測定不確かさという考え方と機械学習の推定結果との関係を明確にすることが重要である．そもそも，精密計測では物理モデルが確立されているため，測定結果を最小二乗法で解析することが容易である．したがって，機械学習に頼らずとも測定結果の確率分布を推定することができる．また，機械学習の推定に関わるアルゴリズムを精密計測に適用する際に，アルゴリズムのブラックボックス化を避けることも重要である．しかしながら，機械学習の特徴や推定アルゴリズムには，従来の精密計測では得られない意義があり，精密計測の分野にとって1つの有効なツールとなりうることがわかる．また，オートエンコーダのようなアルゴリズムは，非線形最小二乗法が発展したものとして捉える

Chapter 13　Machine Learning and Precision Metrology

　As machine learning has become more widely in use, a various new approaches have been developed. In the field of precision metrology, machine learning has been introduced and new progress has been realized. Meanwhile, it is important to clarify, the relationship between the concept of measurement uncertainty, which is considered to substantiate the measurement results in precision metrology, and the estimation results of machine learning. In the first instance, it is easy to analyze measurement results by the least-squares method because a physical model has been established for precision metrology. Therefore, it is possible to estimate the establishment distribution of measurement results without relying on machine learning. In addition, it is also important not to treat the algorithms involved in machine learning estimation as a "black box," when the algorithms are applied for precision metrology. However, the features and estimation algorithms of machine learning have significance that cannot be obtained from conventional precision metrology, and it is clear that machine learning is an effective tool for precision

ともできる．そこで，本章では，機械学習の基本的な考え方を述べた後に，物理的
意義を概説し，実装例を示す．さらに，推定値の不確かさの意味について述べる．

● 13.1　機械学習

機械学習とは，人工知能の研究分野の1つであり，データを解析し，予測や推論を
行うための規則性やルールを見つけ出す方法である．いわゆる人間と同じ知的な処理
能力の確立を目指している人工知能（AI）の一部であることに注意されたい．すなわ
ち，機械学習では，推定値は得られるものの，それ以降何らかの判断をするというこ

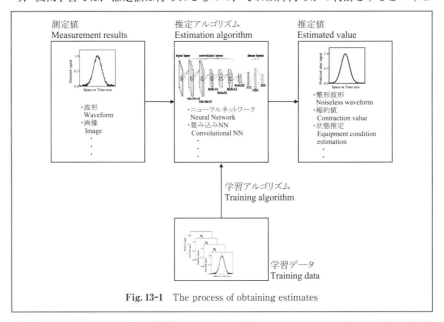

Fig. 13-1　The process of obtaining estimates

metrology. Therefore, in this chapter, after describing the basic concepts of machine learning, we outline its physical significance and present an example implementation. Furthermore, the meaning of the uncertainty of the estimated values is
discussed.

● 13.1　Machine learning

Machine learning is one of the research fields of artificial intelligence, a method of
analyzing data and finding regularities and rules to make predictions and infer-
ences. Note that it is not similar to the so-called AI, which aims to establish the
same intellectual processing capability as humans. That is, in machine learning, estimates are obtained, but no further decisions are made.

とはない.

　機械学習で推定値を得るプロセスは, 一般的に, Fig. 13-1 に示すとおりである. ま ず, 何らかの測定値を得ると, その測定結果を既知の学習データで学習させた推定ア ルゴリズムで解析する. 解析の結果として, 所望の推定値が得られるようにアルゴリ ズムを設計する. 代表的なものとしては, ニューラルネットワークや畳み込みニュー ラルネットワークがあり, 詳細については13.2節で述べる. ここでの推定値は, 例え ば, 入力波形をデノイズさせた波形であったり, 平均値や標準偏差などの縮約値のよ うな定量的な量を設定することができる. なお, 推定アルゴリズム内には, 大量の未 知パラメータが存在しているため, それらのパラメータを最適化するために, 大量の 学習データが必要となる. また, 学習の過程では, 過度に最適化されるオーバーフィッ ティングに注意する必要がある.

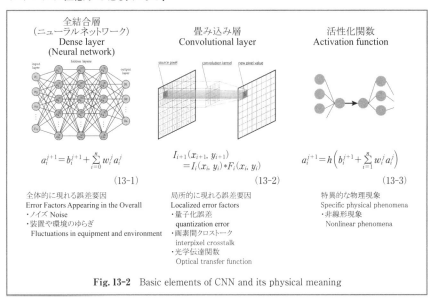

Fig. 13-2　Basic elements of CNN and its physical meaning

　The process of obtaining estimates by machine learning is generally as shown in Fig. 13-1. First, after obtaining some measurements, the measurement results are analyzed by an estimation algorithm trained on known training data. The algorithm is designed so that the result of the analysis yields the desired estimate. The estimated value can be, for example, a denoised waveform of the input waveform, or a quantitative quantity such as a mean value, standard deviation, or other reduced value. Since there are a large number of unknown parameters in the estimation algorithm, optimization is performed on a large amount of training data in order to optimize these parameters.

● 13.2 推定アルゴリズムの物理モデル

機械学習による推定結果は推定アルゴリズムに大きく影響される．この推定アルゴリズムでは，神経細胞の働きを模擬したニューラルネットワーク（NN）が古くから用いられているが，コンピュータ技術の進展や畳み込み層を利用した畳み込みニューラルネットワーク（CNN）の登場により，アルゴリズムの深層化が飛躍的に進み利用価値が広がった．このCNNは，基本的には，全結合層（NN層），畳み込み層（C層）および活性化関数の3つの要素から成り立っている（Fig. 13-2）．NN層やC層では，複数を組み合わせた層を構築することができる．なお，値が入力される入力層，出力の途中の中間層および推定結果の得られる出力層と呼ぶ．

全結合層であるNNでは，入力値のすべてもしくは一部に係数を割り当ててすべてを，式(13.1)で示すように和算することで次の層に値を受け渡す．ただし，和算する際にバイアス b も考慮する．畳み込み層では，ある係数行列を用意し，その行列で畳み込まれた出力値を次の層に伝達する．この畳み込む行列は2次元であることが多いが，複数の2次元行列を用いることもできる．さらに，活性化関数では，各層での出

● 13.2 Physical model on machine learning

Estimation results by machine learning are greatly influenced by the estimation algorithm. Neural networks (NN), which simulate the function of neurons, have long been used in estimation algorithms, but advances in computer technology and the emergence of convolutional neural networks (CNN), which use convolutional layers, have dramatically deepened the depth of the algorithms and expanded their value. The development of computer technology and the appearance of convolutional neural networks (CNN), which use a convolutional layer, have dramatically increased the depth of the algorithms and expanded the value of their use. The CNN is basically composed of three elements: the all-connected layer (NN), the convolutional layer (C), and the activation function (Fig. 13-2). The NN and C can be combined to form multiple layers, which are called the input layer where values are input, the intermediate layer in the middle of the output, and the output layer where the estimation results are obtained.

In an NN, which is an all-coupled layer, all or part of the input values are assigned a count and all of them are summed to pass the value to the next layer as shown in Eq. (13-1). However, the bias b is also taken into account in the summation process. In the convolution layer C, a certain counting matrix is prepared, and the output values that have been collapsed by this matrix are transferred to the next layer. This matrix is often two-dimensional, but can be three-dimensional. The activation func-

力値を受け渡す際に非線形性を持たせることができる.

　CNN における基本要素の特徴を考慮すると各層に物理的な意味を持たせることができる. NN 層での計算は全体に影響をおよぼすと考えると測定系に存在するノイズや装置もしくは環境のゆらぎを表現できる. また, C 層では, 局所的な部分しか次の層に反映されないため, 測定装置の量子化誤差やクロストーク, 光学伝達関数のような物理モデルを割り当てることができる. 活性化関数では, 伝達する値に非線形性を持たせることができるため, 測定装置や原理に内在する非線形性を表現することができる. このように, 一見ブラックボックスに見える推定アルゴリズムでも, その働きを物理的に解釈することが可能である.

● 13.3　機械学習を用いた距離計測

　三角測量に機械学習を適用することで高精度計測を実現した例を Fig. 13-3 に示す. 光源を白色光として, その光をある距離に固定された被測定物の表面にレンズで集光する. 表面で散乱した光は, 観察軸側のレンズで検出器上に集光する. 三角測量であるため, 被測定物の z 方向の位置が, 検出器上で水平方向に変換される. すなわち,

tion allows for nonlinearities in the passing of output values at each layer.

Considering the characteristics of the basic elements in a CNN, each layer can have a physical meaning: the computation in the NN layer can represent the noise in the measurement system and the fluctuations of the device or the environment, considering that they affect the whole system. In the C layer, only a localized part is reflected in the next layer, allowing the assignment of physical models such as the quantization error of the measurement device, crosstalk, or optical transfer function. The activation function allows for nonlinearity in the transmitted values and thus can represent the inherent nonlinearity of the measurement device or principle. Thus, even seemingly black-box estimation algorithms can be given physical meaning.

● 13.3　Displacement measurement with machine learning

Fig. 13-3 shows an example of a highly accurate measurement system that uses a wide range of machine learning techniques for triangulation measurement. A white light source is used to focus the light on the surface of a measured object fixed at a certain distance by a lens. Light waves scattered on the surface are focused onto the detector by a lens on the observation axis side. The position of the object to be measured in the z-direction is converted to the horizontal direction on the detector because of the triangulation technique. In other words, the position of the object to

検出器上での被測定物の位置を算出すればよい．測定システムには，強度や位置にランダムなノイズ成分が存在するため測定誤差が大きくなる．そこで，散乱光の中心位置の推定に機械学習を利用した．推定のアルゴリズムには CNN を用いた．CNN の学習では，学習データを数値計算より求めた．具体的には，理想的な正規分布に，事前に測定で求めたノイズ成分を重畳させた．その結果，精度検定をしたところ，測定器の空間的な分解能である 1pixel 以下の精度で位置を推定できることが示されている．

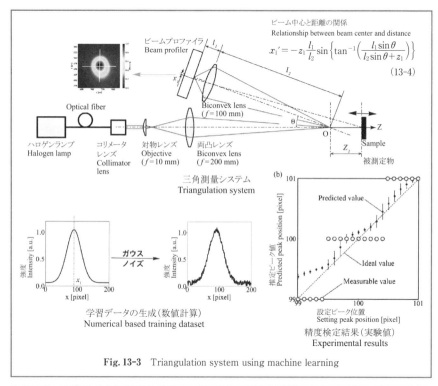

Fig. 13-3 Triangulation system using machine learning

be measured on the detector can be calculated. The presence of noise components in the measurement system, whose intensity and position are random, increases the measurement error. Therefore, machine learning was used to estimate the position of the center of the scattered light. For the CNN training, the training data were obtained from numerical calculations. Specifically, an ideal normal distribution was superimposed on a noise component obtained from prior measurements. Accuracy tests show that the position can be estimated with an accuracy of less than 1 pixel, which is the spatial resolution of the measuring instrument.

● 13.4　機械学習における推定不確かさ

機械学習で得られる推定値の信頼性は，推定不確かさという指標で評価することができる．ここでいう不確かさは，GUM で定義されている不確かさとは異なり，推定値のばらつきのことを指す．Fig. 13-4 に，推定値の分類と推定不確かさの算出方法の例を示す．精密測定における機械学習による推定値は，設定値に対して対象となる物理モデルに即した値となる．測定不確かさが存在するため物理モデル付近でばらつきを持ったモデルとなる．機械学習による推定値は学習データに即した値となるため，測定値のばらつきと同等の範囲で推定値が得られることになる．一方で，学習データと異なる入力値に対しても何らかの推定値が提示されるが物理モデルとは大きく異

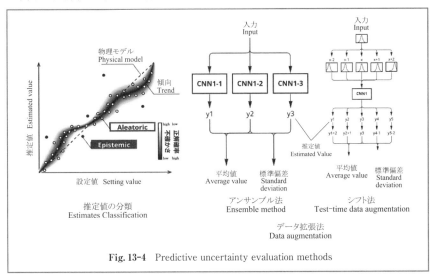

Fig. 13-4　Predictive uncertainty evaluation methods

● 13.4　Predictive uncertainty

The reliability of the estimates obtained by machine learning can be evaluated in terms of the predictive uncertainty (Fig. 13-4). The estimated value by machine learning in precision metrology is the value that corresponds to the physical model of interest for the set value. Because of the existence of measurement uncertainty, the model will have variations around the physical model. Since the estimated value by machine learning is in line with the training data, the estimated value is obtained within the same range as the variation of the measured value. On the other hand, some estimates are provided for input values that differ from the training data. The former is called Aleatoric and the latter is called Epistemic. In this case, values that are Epistemic need to be excluded as outliers because they deviate from the probability distribution of the original data. However, it is extremely difficult to

なった値となる．前者は Aleatoric，後者は Epistemic と呼ばれている．Epistemic と
なった値は，本来のデータが持つ確率分布から外れた値となるため外れ値として除外
する必要がある．一方，Aleatoric となった推定値は，本来の確率分布内の値である．
すなわち，測定に用いることのできる推定値は Aleatoric な値しか採用することがで
きない．しかしながら，1 つの推定値だけで Epistemic と判断することは極めて困難
であり，複数の推定値を得ることでばらつきを評価する必要がある．

　推定値が Aleatoric か否かを判断する方法としてデータ拡張によりデータのばらつ
きを評価する方法が提案されている．アンサンブル法では，学習アルゴリズムは同じ
であるものの，初期値をわずかに変えて学習させた複数のアルゴリズムを用意し，各
アルゴリズムに対して同じ測定値を入力させる．その結果，アルゴリズムの数だけ推
定値が得られるが，入力値に異常値などが含まれなければ，各アルゴリズムの標準偏
差が小さくなる．一方で，異常値が含まれると標準偏差が大きくなる．他には，シフ
ト法という方法がある．推定アルゴリズムは 1 つであるが，測定値を意図的にずらし
たデータを複数用意する．その後，それらのデータを入力し，各々のデータによる推
定結果を入力時にずらした分だけ元に戻す．最終的にはアンサンブル法と同様に推定
値を評価することができる．このシフト法は，学習プロセスが 1 回で済むというメリッ

determine an outlier with only one estimate.

A method to evaluate data variability by data expansion has been proposed as a
method to determine if an estimate is Aleatoric or not. In the ensemble method,
multiple algorithms are trained with the same learning algorithm but with slightly
different initial values, and the same measurement values are input to each algorithm. As a result, estimates are obtained for each algorithm, but if the input values
do not include anomalies, the standard deviation for each algorithm becomes small.
On the other hand, if abnormal values are included, the standard deviation becomes
large. Another method is the shift method, which uses a single estimation algorithm
but prepares multiple sets of data with intentionally shifted measured values. The
data are then input, and each data is restored to its original value by the amount
that was shifted at the time of input. Estimates can then be evaluated in the same
way as with the ensemble method. This shifting method has the advantage of requiring only one learning process.

Fig. 13-5 shows the uncertainty factors of the estimate from the perspective of
precision metrology. To obtain the final estimate, there are two main processes: the
measurement process by experiment and the estimation process by the estimation

トがある.

　推定値の採否を決める要因を精密計測の視点から分析したのが Fig. 13-5 である. 最終的な推定値を得るためには, 大きく分けて, 実験による測定プロセスと推定アルゴリズムによる推定プロセスがある. 測定値は, 測定器が存在する環境の状態に依存する. そのため, 学習データと異なる条件になると推定値は Epistemic になる.

　また, 環境状態が変化していないとしても, 測定値にはばらつきが存在する. しかし, このばらつきは, 学習データに含まれている内容であるので Aleatric の範囲の推定値が得られる. 一方で, 推定値は, 推定アルゴリズムおよび学習に依存する. 物理モデルを正確に表現できていない推定モデルを用いると推定値のばらつきは大きくなる. また, 学習が十分に行われなければ推定値のばらつきは同様に大きくなり Epistemic となってしまう. 他にも, 測定データに何らかの異常値が混入した場合でも推定値を得られることになるが Epistemic になることは自明である. このように, 測定システムに機械学習を導入する際には, これらの誤差要因が含まれているので, 推定不確かさの評価が必要不可欠である. また, 推定不確かさが測定環境に依存していることを考えると, 推定不確かさを評価することで測定システムのキャリブレーション

algorithm. The measured values depend on the conditions of the environment in which the measuring instrument exists. Therefore, if the conditions are different from the training data, the estimates become Epistemic. Even if the environmental conditions have not changed, there is still variability in the measured values. However, since this variation is contained in the training data, the estimated values are within the Aleatric range. On the other hand, the estimates depend on the estimation algorithm and the training. If an estimation model that does not accurately reflect the physical model is used, the variability of the estimates will be large. Even if the estimation algorithm and the training data are correct, if the training is not done sufficiently, the variability of the estimates will increase as well and become Epistemic. It is also obvious that even if some anomalous values are mixed in the measurement data, the estimated values will be obtained, but they will be Epistemic. Thus, when introducing machine learning into a measurement system, it is essential to evaluate the predictive uncertainty since these error factors are included. Moreover, considering that the predictive uncertainty depends on the measurement environment, it is also useful to evaluate the predictive uncertainty so that it can be used for system evaluation, such as determining the calibration timing of the measurement system. In addition, it is important to follow GUM when making uncertainty analysis on the data processing results in precision measurement where ma-

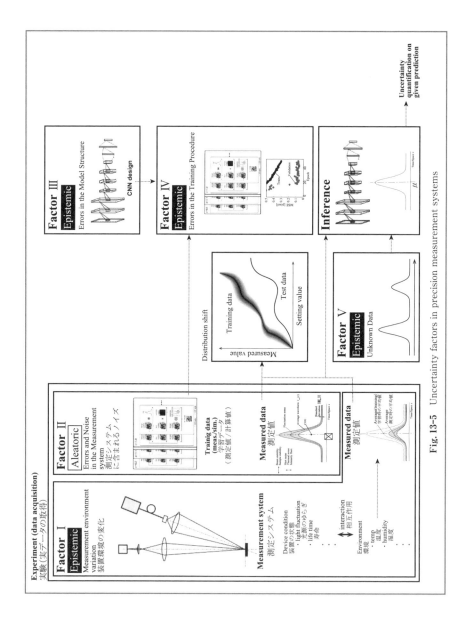

Fig. 13-5 Uncertainty factors in precision measurement systems

タイミングの見極めなどシステム評価に利用できる点も有用である．一方，機械学習が導入された精密計測において，データ処理結果の不確かさ解析を GUM に従って行うことが重要である．

【演習問題】

13-1) 推定アルゴリズムの1つにオートエンコーダという方法があるが，その特徴について調査し，最小二乗法によるフィッティングとの違いについて述べよ．

13-2) 推定アルゴリズムのうち，U-Net や ResNet が報告されているが，それらを調査し，それぞれの特徴を物理的な解釈を踏まえて説明せよ．

13-3) 活性化関数が線形変換の場合に生じるアルゴリズムの問題点について述べよ．

13-4) 学習プロセスにおいて学習しすぎたときに生じうる問題について述べよ．

chine learning is applied. In addition, it is important to follow GUM when making uncertainty analysis on the data processing results in precision measurement where machine learning is applied.

【Problems】

13-1) Investigate the characteristics of autoencoder, one of the estimation algorithms, and describe how it differs from least-squares fitting.

13-2) Among the estimation algorithms, U-Net and ResNet have been reported.

13-3) Describe the algorithmic problems that arise when the activation function is a linear transformation.

13-4) Describe problems that can arise when too much learning occurs in the learning process.

第14章　超短パルスレーザと光周波数コム

　光源は光計測にとって欠かせない要素である．光学式センサで用いられる光源は，
1）発光ダイオード（LED）で代表される白色光源と，2）半導体レーザ（LD）で代表
されるレーザ光源，に大別できる．前者はブロードな光周波数スペクトルを持つ低コ
ヒーレンス光であるのに対して，後者は高い指向性を持つコヒーレント光である．ま
た，後者には連続波（CW）の光を出力する単一周波数レーザと，Fig.14-1 に示すよ
うなパルス状の光を出力するパルスレーザがある．特にパルス幅がフェムト秒程度の
超短パルスレーザは，白色光源のようにブロードなスペクトルを持ち，白色レーザと
呼ばれる．本章では超短パルスレーザとその光スペクトルを安定化した光周波数コム
について述べた後，その精密計測における応用を紹介する．

Chapter 14　Ultrashort Pulse Laser and Optical Frequency Comb

　Light sources are an essential element in optical metrology. Such a light source
can be categorized as 1) a white light source such as LED that generates an incoherent light with a broad optical spectrum and 2) a laser source such as the laser
diode that generates a coherent light with a high directivity. Laser sources can be
further categorized into continuous wave (CW) and pulse laser sources. A CW laser outputs a monochromatic light beam at a constant intensity and a single frequency. A pulse laser outputs a train of light pulses, as shown in Fig. 14-1. An ultrashort pulse laser with a pulse width in the order of fs, often called a white light
laser, has a broad spectrum comparable to that of a white light source. This chapter
presents the ultrashort pulse laser and its stabilized spectrum (optical frequency
comb), as well as their applications in precision metrology.

● **14.1 超短パルスと光周波数コムの生成**

Fig. 14-1 に長さが L_{cav} で損失や分散などを考慮しない理想的なファブリ・ペロー共振器を示す．共振器は 2 枚のミラーから構成される．左側のミラーは高い反射率を持ち，右側のミラーは光が透過できるように部分反射するようになっている．光が共振器のミラー間で 1 往復する時間を $\tau_r(=1/f_r)$ とすると，Fig. 14-2 に示すように，一定の周波数間隔 $\omega_r(=2\pi f_r)$ を持つ縦モードの配列が生成される．また Fig. 14-3 のように，すべての縦モードの初期位相は点 O において 0 である．ここで，レーザ媒質の利得は周波数 $\omega_c(=2\pi f_c)$ で最大値を取り，それを中心に対称となるとする．ω_c を持つモードを中心モードと呼び，その電場を $E_0(t)(=a_0 E_c(t))$ で表す．ここで，a_0 は電場振幅であり，$E_c(t)(=e^{j\omega_c t})$ は搬送波と呼ばれる．中心モード電場 $E_0(t)$ に，同じ電場振幅 a_m を持つ $\pm m$ モード対の電場 $E_{\pm m}(t)$ との和を取ると，搬送波の電場振幅に正弦波的変調が加えられた合成波が得られる．変調された振幅関数 $g(t)$ は合成波の包絡線と

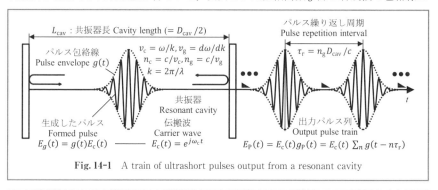

Fig. 14-1   A train of ultrashort pulses output from a resonant cavity

● **14.1   Generation of ultrashort pulses and optical frequency comb**

Fig. 14-1 shows a laser oscillator of an ideal Fabry-Pérot resonant cavity with a cavity length of L_{cav}, in which no losses and dispersion are considered. The cavity comprises two mirrors, one on the left with a high reflectivity and the other on the right with partial transmission, for surrounding the laser gain medium. As shown in Fig. 14-2, a series of cavity longitudinal modes, which are standing wave harmonics with a constant frequency spacing of ω_r or $2\pi f_r$ and an initial phase angle of zero at point O shown in Fig. 14-3, are generated in the cavity as the light propagates forward and backward between the two mirrors with a round-trip time of $\tau_r(=1/f_r)$. Assume the laser gain medium has a symmetric spectrum centering at frequency ω_c or $2\pi f_c$, where the gain is the maximum and the $\pm m$th modes have the same amplitude a_m. The mode at ω_c is referred to as the central mode $E_0(t)(=a_0 E_0(t))$, which is the product of an amplitude a_0 and a carrier wave $E_c(t)(=e^{j\omega_c t})$. Adding such a pair of $\pm m$th modes to the central mode results in a sinusoidal modulation of

してみることができる．中心モードの電場にさらに多くのモード対の電場が加えられると，合成波はピーク位置が点 O のパルス状に近付いて行く．すべての縦モードを足し合わせた結果，Fig. 14-1 に示すように，電場が $E_g(t)$（$=g(t)E_c(t)$）で表されるパルスが共振器内で生成される．それに伴い，時間間隔が τ_r の光パルス列が共振器から出力される．ここで，τ_r はパルス繰り返し間隔と呼ばれ，群屈折率 n_g，共振器長 L_{cav} および真空中の光速 c で決定される．τ_r の逆数である f_r（$=1/\tau_r$）はパルス繰り返し周波数と呼ばれる．光パルス列の電場 $E_P(t)$ はパルス包絡線列 $g_P(t)$（$=\sum_n g(t-n\tau_r)$）と搬送波電場 $E_c(t)$ の積となる．ここでは，レーザ利得媒質は分散がないとすると，群屈折率 n_g は位相屈折率 n_c と，またパルス包絡線の群速度 v_g は搬送波 $E_c(t)$ の位相速度 v_c とそれぞれ同じとなる．

Fig. 14-4(a) に示すように，$g(t)$ をガウス関数と仮定すると，そのフーリエ変換

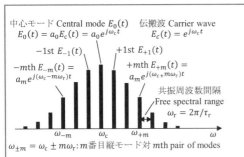

中心モード Central mode $E_0(t)$　伝搬波 Carrier wave
$E_0(t) = a_0 E_c(t) = a_0 e^{j\omega_c t}$　　$E_c(t) = e^{j\omega_c t}$

$-$1st $E_{-1}(t)$,　　$+$1st $E_{+1}(t)$

$-m$th $E_{-m}(t) =$　　　$+m$th $E_{+m}(t) =$
$a_m e^{j(\omega_c - m\omega_r)t}$　　　$a_m e^{j(\omega_c + m\omega_r)t}$

共振周波数間隔
Free spectral range
$\omega_r = 2\pi/\tau_r$

ω_{-m}　　ω_c　　ω_{+m}　　ω

$\omega_{\pm m} = \omega_c \pm m\omega_r$：$m$番目縦モード対 mth pair of modes

Fig. 14-2　Pairs of longitudinal modes symmetrical about the central mode

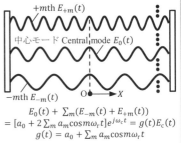

$+m$th $E_{+m}(t)$

中心モード Central mode $E_0(t)$

$-m$th $E_{-m}(t)$　　O　　X

$E_0(t) + \sum_m (E_{-m}(t) + E_{+m}(t))$
$= [a_0 + 2\sum_m a_m \cos m\omega_r t]e^{j\omega_c t} = g(t)E_c(t)$
$g(t) = a_0 + \sum_m a_m \cos m\omega_r t$

Fig. 14-3　Summation of the longitudinal modes with the same initial phase angle at point O

the amplitude of the carrier wave with a modulated amplitude function $g(t)$. Adding more pairs of modes makes $g(t)$ to approach to a pulsed shape with its peak at point O. Eventually, a pulse with an electrical field $E_g(t)$ ($=g(t)\,E_c(t)$) is formed in the cavity from a superposition of all the modes, and a pulse train with a time period of τ_r are then output from the cavity as shown in Fig. 14-1, where $g(t)$ is called the pulse envelope. τ_r is called the pulse repetition interval, which is determined by n_g, L_{cav}, and the speed of light c in vacuum. f_r($=1/\tau_r$) is referred to as the repetition frequency. The electric field $E_P(t)$ of the pulse train is the product of $g_p(t)$ ($=\sum_n g(t-n\tau_r)$) and $E_c(t)$. Without considering the dispersion of the laser medium, the group refractive index n_g is equal to the phase refractive index n_c, and the group velocity v_g of the pulse envelope $g(t)$ is equal to the phase velocity v_c of the carrier wave $E_c(t)$.

Assuming $g(t)$ in Fig. 14-4(a) is a Gaussian function, its Fourier transform $G(\omega)$

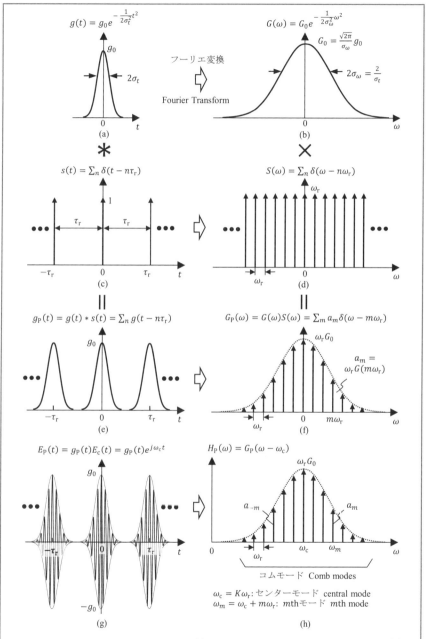

Fig. 14-4 Analysis of the electric field $E_P(t)$ of pulse train and its Fourier transform $H_P(\omega)$ (optical frequency comb) in an ideal resonant cavity without dispersion

$G(\omega)$ もガウス関数となる．$G(\omega)$ は Fig. 14-4(b) に示す．図中の σ_t と $\sigma_\omega (=1/\sigma_t)$ はそれぞれ $g(t)$ と $G(\omega)$ の RMS 幅（2乗平均平方根幅）であり，ここでは便宜的に，FWHM 幅（半値全幅）の代わりに用いられている．g_0 と G_0 はそれぞれ $g(t)$ と $G(\omega)$ の $t=0$ における値である．$s(t)$ を，Fig. 14-4(c) に示すような時間間隔が τ_r のインパルス関数の系列とすると，$s(t)$ のフーリエ変換 $S(\omega)$ は Fig. 14-4(d) に示すような周波数間隔 ω_r のインパルス列となる．

パルス包絡線列 $g_P(t)$ はパルス包絡線関数 $g(t)$ を時間間隔 τ_r で等間隔に並べた系列である．Fig. 14-4(e) に示すように，$g_P(t)$ は数学的に $g(t)$ と $s(t)$ の畳み込み積分と一致する．また，Fig. 14-4(f) に示すように，$g_P(t)$ のフーリエ変換 $G_P(\omega)$ は $G(\omega)$ と $S(\omega)$ の積となる．これは $S(\omega)$ によって $G(\omega)$ が周波数間隔 ω_r でサンプリングされることに相当する．サンプリングして得られた $G_P(\omega)$ は等間隔のインパルス関数からなる．$G_P(\omega)$ の m 番目のインパルスは周波数 $m\omega_r$ を持ち，m 番目インパルスモードと呼ばれる．Fig. 14-4(f) に示すように，$G_P(\omega)$ の m 番目のモードの振幅は a_m である．

Fig. 14-1, 14-4(g) に示される $E_P(t)$ は $g_P(t)$ と $E_c(t)$ の積であるため，周波数シフト効果に基づき，$E_P(t)$ のフーリエ変換 $H_P(\omega)$ は Fig. 14-4(h) のように求められる．

will also be a Gaussian function, which can be expressed in Fig. 14-4(b). Here, σ_t and $\sigma_\omega(=1/\sigma_t)$ are the root-mean-square (RMS) widths of $g(t)$ and $G(\omega)$, respectively. For the sake of clarity, the RMS width is employed here instead of the FWHM width. g_0 and G_0 are the values at $t=0$. Let $s(t)$ in Fig. 14-4(c) be an infinite sequence of impulse functions with a time interval of τ_r. The Fourier transform $S(\omega)$ of $s(t)$ is also an infinite sequence of impulse functions with an angular frequency interval of ω_r (Fig. 14-4(d)).

The pulse envelope train $g_P(t)$, which is a series of pulse envelope functions $g(t)$ equally-spaced with a time period τ_r, can be mathematically produced from the convolution of $g(t)$ and $s(t)$, while the Fourier transform $G_P(\omega)$ of $g_P(t)$ can be obtained from the product of $G(\omega)$ and $S(\omega)$, as shown in Fig. 14-4(e) and (f), respectively. Here, $S(\omega)$ works as a sampling function to sample $G(\omega)$ with a frequency interval of ω_r. The sampled $G_P(\omega)$ is a discrete function comprising a sequence of equidistant impulses. The mth impulse of $G_P(\omega)$ with an angular frequency $m\omega_r$ is referred to as the mth mode. The amplitude of the mth impulse mode in $G_P(\omega)$ is a_m, as shown in Fig. 14-4(f).

Since $E_P(t)$ in Fig. 14-1 and 14-4(g) is the product of $g_P(t)$ and $E_c(t)$, the Fourier transform $H_P(\omega)$ of $E_P(t)$ can be obtained in Fig. 14-4(h) based on the frequency-shift effect. $H_P(\omega)$ is referred to as the optical frequency comb, comprising

$H_P(\omega)$ は光周波数コムと呼ばれる．この光周波数コムは包絡線関数 $G(\omega-\omega_c)$ を持ち，等間隔の離散コムモードから構成される．Fig. 14-4(h) にある ω_m は $H_P(\omega)$ の m 番目のインパルスモード（コムモード）の角周波数である．m 番目コムモードの角周波数 ω_m は Fig. 14-4(f) に示され $G_P(\omega)$ の m 番目インパルスモードの周波数 $m\omega_r$ から ω_c だけシフトしている．また，$H_P(\omega)$ の m 番目コムモードと $G_P(\omega)$ の m 番目インパルスモードの振幅はともに a_m となっている．周波数軸上におけるそれぞれのコムモードの位置は中心周波数 ω_c とコム間隔 ω_r で表すことができる．時間間隔 τ_r（あるいは $2\pi/\omega_r$）が，搬送波の K 周期分に等しいとき，ω_c と ω_r は Fig. 14-4(h) に示されているような関係がある（$\omega_c=K\omega_r$）．光周波数コム $H_P(\omega)$ は ω_c を中心に対称的な包絡線関数 $G(\omega-\omega_c)$ を持つとすると，$\pm m$ 番コムモードの角周波数が $\omega_{\pm m}$ と表され（$\omega_{\pm m}=\omega_c\pm m\omega_r$），その振幅が同じ a_m となる．

一方，パルス列は物理的にレーザ共振器の縦モードの足し合わせで形成されている．分散がない理想的な共振器の場合，Fig. 14-4(h) の光周波数コム $H_P(\omega)$ は Fig. 14-2 に示す共振器中のレーザ媒質の利得スペクトルに相当する．利得スペクトルの中心周波数は搬送波の周波数，あるいは光周波数コムの中心周波数に対応し，光周波数コムの

equally spaced discrete comb modes with a comb envelope of $G(\omega-\omega_c)$. Here, ω_m is the angular frequency of the mth impulse mode (mth comb mode). The angular frequency of the mth comb mode in $H_P(\omega)$ of Fig. 14-4(h) has a frequency shift of ω_c from that of the mth impulse mode in $G_P(\omega)$ of Fig. 14-4(f), while the amplitude is kept at the same a_m. The position of each comb mode on the frequency axis (i.e., mode position) is determined by that of the central mode ω_c and the comb spacing ω_r. Assuming K periods of the carrier wave exist during the time period of τ_r (or $2\pi/\omega_r$), the relationship between ω_c and ω_r is expressed in Fig. 14-4(h). The angular frequencies of the pairs of the $\pm m$th comb modes in $H_P(\omega)$ can be expressed by $\omega_{\pm m}$, of which amplitudes are the same of a_m by assuming the comb has a symmetric envelope of $G(\omega-\omega_c)$ about ω_c.

However, the train of pulses is physically formed by the superposition of the longitudinal modes in a laser resonant cavity. In the case of an ideal resonant cavity without dispersion, the optical frequency comb $H_P(\omega)$ in Fig. 14-4(h) corresponds to the lasing spectrum of the laser gain medium in the resonant cavity (Fig. 14-2), where the central frequency of the gain curve of the laser gain medium corresponds to the carrier frequency ω_c (i.e., the frequency of the central mode of the optical frequency comb). The comb width $2\sigma_\omega$ corresponds to the spectral bandwidth of the laser gain medium. The pulse width $2\sigma_t$ shown in Fig. 14-4(a) is referred to

幅 $2\sigma_\omega$ はレーザ媒質の周波数利得幅に対応することとなる．また，Fig. 14-4(a)に示される パルス幅 $2\sigma_t$ はフーリエ変換限界パルス幅，あるいは周波数帯域限界パルス幅と呼ばれ，レーザ媒質の利得スペクトルの周波数帯域幅に決定される．周波数帯域幅が広ければ広いほど，パルス幅が狭くなる．例えば，σ_ω が 10 THz の場合，σ_t は 100 fs 程度となる．この程度のパルス幅を持つモードロックレーザはフェムト秒レーザと呼ばれる．実際の共振器の場合，レーザ媒質の屈折率は屈折率分散の影響により，周波数 ω によって変化してしまう．そのために，パルス幅は周波数帯域限界パルス幅より広くなってしまう．より重要なことに，共振器縦モード間の周波数間隔は一定でなくなり，また，縦モードの初期位相も等しくならなくなる．フーリエ変換の理論により，そのような非等間隔配列と不等初期位相の共振器縦モード和は光パルスを形成することができない．そこで，モードロック技術を用いて分散の影響を取り除くことで，光パルス形成に必要な等初期位相を持つ等ピッチ縦モードを実現する．

as the width of the Fourier transform-limited pulse or the bandwidth-limited pulse, which is inversely proportional to the spectral bandwidth. A wider spectral bandwidth results in a shorter the pulse width. For σ_ω of 10 THz, σ_t is calculated to be 100 fs. A mode-locked laser with such a short pulse width is called a femtosecond laser. The width of a practical pulse exceeds that of the bandwidth limited pulse in an actual resonant cavity where the refractive index of the laser gain medium changes with frequency ω, which is called dispersion of refractive index, or simply dispersion. Notably, the frequency separation between the two adjacent cavity longitudinal modes of such a resonant cavity is no longer constant. The initial phase angles of the longitudinal modes are also not equal with each other. Based on the theory of Fourier transform, an optical pulse in the resonant cavity cannot be formed from the superposition of such longitudinal modes with nonconstant frequency spacing and/or unequal initial phase angles. A mode-locking technique is therefore required for both the frequency separation and the initial phase angle to be constant so that an optical pulse can be formed without the influence of dispersion.

● **14.2 モードロックの原理**

振幅変調に基づくモードロックの原理を Fig. 14-5 に示す. 中心縦モードの振幅を変調し, 一対のサイドバンドモード $\widetilde{E}_{\pm 1}(t)$ を形成することが示されている. そのために必要な変調器として, 電気光学素子や音響光学素子による能動タイプのものと, 電気光学カー効果に基づく可飽和吸収体を用いた受動タイプのものがある. 変調器は共振器内に設置し, 共振器の損失に周期的な変調をかけるようにする. その結果として, 共振器中の縦モードは角周波数が ω_r, 振幅が $b(t)$ の正弦関数 $E_{\mathrm{mod}}(t)$ で変調されることとなる (Fig. 14-5(a)). $E_{\mathrm{mod}}(t)$ の時間周期 τ_r は縦モードの合成波の共振器内を 1 往復する時間に近い値に設定する. τ_r は群速度 v_g で決まる. $b(t)$ のフーリエ変換を $B(\omega)$ とし, ここで $b(t)$ は緩やかに変化する関数とする. $B(\omega)$ の周波数帯域は ω_r より低いため, 中心縦モード電場 $E_0(t)$ の振幅 a_0 は $E_{\mathrm{mod}}(t)a_0$ に変調される. 変調された電場 $E_{\mathrm{mod}}(t)E_0(t)$ のスペクトルは畳み込み積分 $E_{\mathrm{mod}}(\omega)*E_0(\omega)$ の逆フーリエ変換になり, 二つのサイドバンドモード $\widetilde{E}_{-1}(\omega)$, $\widetilde{E}_{+1}(\omega)$ が生成される. Fig. 14-5(b)に示すように, $\widetilde{E}_{-1}(\omega)$ と $\widetilde{E}_{+1}(\omega)$ は周波数域において中心モードの左右に等間隔に配置されている. $\widetilde{E}_{-1}(\omega)$ と $\widetilde{E}_{+1}(\omega)$ の初期位相は中心モードの位相 ϵ_0 と同じとなる. モード引込効

● **14.2 Principle of mode-locking**

The principle of mode-locking based on amplitude modulation is schematically demonstrated in Fig. 14-5. Modulation of the amplitude of the central longitudinal mode is carried out for generating a pair of side bands $\widetilde{E}_{\pm 1}(t)$. A modulator, which can be either an electro-optic/acousto-optic modulator for active mode-locking, or a saturable absorber based on the Kerr effect for passive mode-locking, is placed in the resonant cavity to provide periodic modulation of the resonator losses, so that the electric fields of the longitudinal modes in the cavity is modulated with a sinusoidal function of $E_{\mathrm{mod}}(t)$, where ω_r is the angular frequency of $E_{\mathrm{mod}}(t)$, and $b(t)$ is the amplitude (Fig. 14-5(a)). The corresponding time period τ_r of $E_{\mathrm{mod}}(t)$ is set to be close to the round-trip time determined by the group velocity of the superposed wave of the longitudinal modes in the cavity. The Fourier transform of $b(t)$ is denoted by $B(\omega)$. Here, $b(t)$ is a slowly varied function and $B(\omega)$ has a frequency bandwidth much less than ω_r. Since ω_r is much smaller than the frequency ω_c, the amplitude a_0 of the electrical field $E_0(t)$ of the central longitudinal mode is modulated to $E_{\mathrm{mod}}(t)a_0$. The spectrum of the modulated electrical field $E_{\mathrm{mod}}(t)E_0(t)$, which is an inverse Fourier transform of the convolution $E_{\mathrm{mod}}(\omega)*E_0(\omega)$, comprises two side bands $\widetilde{E}_{-1}(\omega)$ and $\widetilde{E}_{+1}(\omega)$. The two side bands produced are equally spaced about the central mode on the frequency domain, as shown in Fig. 14-5(b). The initial

果によって，1番目の縦モード対 $E_{\pm 1}$ はそれぞれのサイドバンドモードにロックされることとなる．その結果として，$E_{\pm 1}$ は中心モードから周波数間隔 ω_{r} で再配置され，また，$E_{\pm 1}$ の初期位相は中心モードの位相 ϵ_0 に揃えられる．再配置された1番目の縦モードはまた同じように2番目の縦モードをロックする．この過程が繰り返されていき，最終的にはすべての縦モードが同じ周波数間隔 ω_{r} と初期位相角 ϵ_0 にロックされるようになる．ロックされたこれらの縦モードの合成で光パルスが生成される．それに対応する光周波数コムの m 番目モードの周波数は $\omega_m = (K+m)\omega_{\mathrm{r}} + \omega_{\mathrm{ceo}}$ と表され，ω_{ceo} はキャリア-エンベロープオフセット周波数と呼ばれる（K は Fig. 14-4(h) 参照）．

● 14.3　精密計測への応用

超短パルスレーザの精密計測への応用は，3つのカテゴリーに分けることができる．

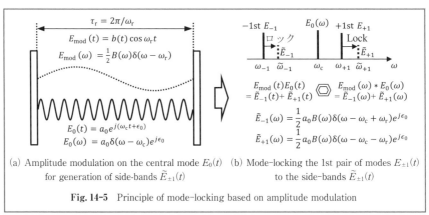

(a) Amplitude modulation on the central mode $E_0(t)$　(b) Mode-locking the 1st pair of modes $E_{\pm 1}(t)$
　　for generation of side-bands $\tilde{E}_{\pm 1}(t)$　　　　　　　　　　to the side-bands $\tilde{E}_{\pm 1}(t)$

Fig. 14-5　Principle of mode-locking based on amplitude modulation

phase angles of $\tilde{E}_{-1}(\omega)$ and $\tilde{E}_{+1}(\omega)$ are equal to that of the central mode ϵ_0. The first pair of longitudinal modes $E_{\pm 1}$ are therefore locked to each of the side bands due to the mode-pulling effect. Consequently, $E_{\pm 1}$ are realigned with an equal spacing ω_{r} and the same initial phase angle ϵ_0 as that of the central mode. The realigned first pair of modes then lock the second pair of modes in the same way. This process is repeated until all modes are locked with the same frequency spacing ω_{r} and the same initial phase angle ϵ_0. Thus, an optical pulse can be generated from the superposition of the locked longitudinal modes. The mode ω_m of the associated optical frequency comb can be expressed by $\omega_m = (K+m)\omega_{\mathrm{r}} + \omega_{\mathrm{ceo}}$, where ω_{ceo} is called the carrier-envelope offset frequency (For K, see Fig. 14-4(h)).

● 14.3　Applications to precision metrology

Applications of ultrashort pulses in precision metrology can be classified into

1つ目は高い安定性と狭い線幅のコムモードを持ち，時間の国家標準に直接リンクできる光周波数コムを利用するものである．2つ目は種々の非線形光学現象を生成できる超短パルスの高いピークパワーを利用するものである．Fig. 14-6 に光周波数コムの m 番目モードを利用して，電場が $E_{CW}(t)$ で表される CW レーザの光周波数 ω_{CW} を測定する原理を示す．ここで光周波数コムの m 番目モード $E_m(t)$ の周波数 ω_m の値が ω_{CW} に 1 番近いとなっている．図に示すように，ω_{CW} と ω_m の差である $\Delta\omega_{CW}$ はビート周波数と呼ばれ，$E_{CW}(t)$ と $E_m(t)$ の干渉信号 $I(t)$ から測定することができる．$\Delta\omega_{CW}$ と ω_m から算出される ω_{CW} は干渉計による精密測長などに用いることができる．Fig. 14-7 は光周波数コムの全スペクトルを利用した角度計測の原理を示す．光周波数コムが格子ピッチ p の回折格子反射鏡に入射すると，1 次回折光群が生成される．m 番目モードによる 1 次回折光の強度 $I(\omega_m, \theta)$ は反射鏡の角度 θ によって変化する．その 1 次回折光の回折角 β_m は $I(\omega_m, \theta)$ のピッチ位置から検出できる．また，反射鏡の角度変位も

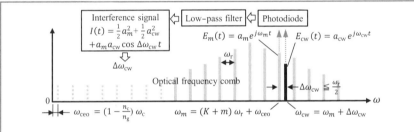

Fig. 14-6   Measurement of optical frequency ω_{CW} of a CW laser by using optical frequency comb

three categories. The first one utilizing the optical frequency spectrum with stable frequency comb modes of narrow linewidth that can be directly linked to national time standards, and the second one utilizing the high peak power of an ultrashort pulse that can create phenomena of nonlinear optics. Fig. 14-6 shows the principle for measuring the optical frequency ω_{CW} of a CW laser with an electric field $E_{CW}(t)$ using the mth comb mode. Assume that the frequency ω_m of the mth comb mode $E_m(t)$ is the closest to ω_{CW}. The difference $\Delta\omega_{CW}$ between ω_{CW} and ω_m, which is referred to as the beat frequency, can be obtained from the interference signal $I(t)$ of $E_{CW}(t)$ and $E_m(t)$, as shown in the figure. The measured $\Delta\omega_{CW}$ and ω_{CW} can then be employed in interferometers for accurate length measurement. Fig. 14-7 shows the principle of angle measurement using the entire spectrum range of the optical frequency comb. When the comb is projected onto a grating reflector with a grating pitch p, the comb modes generate a group of first-order diffracted beams. The intensity $I(\omega_m, \theta)$ of the first-order diffracted beam by the mth mode changes with the angular position θ of the reflector, where the diffraction angle β_m can be identi-

$I(\omega_m, \theta)$ に基づいて測定できる．Fig. 14-8 には非線形光学現象の一種である第二高調波発生を利用して反射ミラーの直動と角度変位を測定するセンサの原理を示す．非線形光学素子によって，基本波である超短パルスレーザの中心周波数 ω_c の2倍の周波数を持つ光が生成され，ミラーの直動と角度変位の検出に利用されている．また，高い指向性，長いコヒーレンス長，ブロードな周波数スペクトルの特徴を持つ白色レーザ光源として長距離分光干渉計などに利用されるのは3つ目の応用カテゴリーとなる．

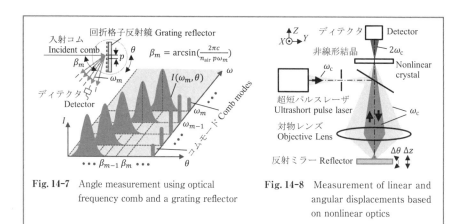

Fig. 14-7 Angle measurement using optical frequency comb and a grating reflector

Fig. 14-8 Measurement of linear and angular displacements based on nonlinear optics

fied from the peak position of $I(\omega_m, \theta)$. Angular displacement of the reflector can also be evaluated from $I(\omega_m, \theta)$. Fig. 14-8 shows a sensor that can detect the linear and angular displacements of a reflector utilizing the nonlinear optics phenomenon of a second harmonic wave with a frequency $2\omega_c$, which is double the central frequency of the ultrashort pulse laser as the primary wave. As the third category of applications, an ultrashort pulse laser can be employed as a white light laser source in long-distance spectral interferometry, etc., where the features of the ultrashort laser source in terms of high directivity, long coherence length and broad frequency spectrum are utilized.

【演習問題】

14-1） Fig. 14-1 の τ_r が 10 ns のときの D_{cav} を計算せよ．ただし，n_g を 1.5 とする．

14-2） Fig. 14-4 において，σ_t を 40 fs としたときの σ_ω を計算せよ．

14-3） 大気中で伝搬する超短パルスレーザの中心波長 λ_c を 1560 nm とし，波長帯域が 1540 nm から 1580 nm までになっているとすると，中心周波数 f_c および周波数帯域を算出せよ．ただし，大気の屈折率は 1 とする．

14-4） 位相速度 v_c と群速度 v_g を周波数 ω と波数 k を用いてそれぞれ表せ．ただし，λ は波長であり，波数 $k=\dfrac{2\pi}{\lambda}$ である．

14-5） Fig. 14-4(h) にある式 $H_P(\omega)=G_P(\omega-\omega_c)$ を証明せよ．

【Problems】

14-1） Calculate D_{cav} in Fig. 14-1 when τ_r is 10 ns. Assume n_g is 1.5.

14-2） Assuming σ_t in Fig. 14-4 is 40 fs, calculate σ_ω.

14-3） For an ultrashort pulse laser propagating in air with a central wavelength λ_c of 1560 nm, and a wavelength bandwidth from 1540 nm to 1580 nm, calculate the central frequency f_c and the frequency bandwidth. Assume that the refractive index of air is 1.

14-4） Express the phase velocity v_c and the group velocity v_g in terms of frequency ω and wave number k, respectively. $k=\dfrac{2\pi}{\lambda}$, where λ is the wavelength

14-5） Prove the equation $H_P(\omega)=G_P(\omega-\omega_c)$ in Fig. 14-4(h).

参考文献
References

1. JIS B 0021：製品の幾何特性仕様（GPS）―幾何公差表示方式―形状，姿勢，位置及び振れの公差表示方式

2. JIS B 0031/Amendment 1：製品の幾何特性仕様（GPS）―表面性状の図示方法

3. JIS B 0405：普通公差―第1部：個々に公差の指示がない長さ寸法及び角度寸法に対する公差

4. JIS B 0420-1：製品の幾何特性仕様（GPS）―寸法の公差表示方式―第1部：長さに関わるサイズ

5. JIS B 0420-2：製品の幾何特性仕様（GPS）―寸法の公差表示方式―第2部：長さ又は角度に関わるサイズ以外の寸法

6. JIS B 0651：製品の幾何特性仕様（GPS）―表面性状：輪郭曲線方式―触針式表面粗さ測定機の特性

7. JIS B 0681-1：製品の幾何特性仕様（GPS）―表面性状：三次元―第1部：表面性状の図示方法

8. JIS B 7516：金属製直尺

9. JIS B 7523：サインバー

10. JIS B 7534：角度直尺

11. JIS B 7541：標準尺

12. JIS Z 8103：計測用語

13. JIS Z 8317-1：製図―寸法及び公差の記入方法―第1部：一般原則

14. ISO 1：Geometrical Product Specifications（GPS）―Standard Reference Temperature for the Specification of Geometrical and Dimensional Properties

15. ISO 1101：Geometrical Product Specifications（GPS）―Geometrical Tolerancing―Tolerances of Form, Orientation, Location and Run-out

16. ISO/DIS 1101：Geometrical Product Specifications（GPS）―Geometrical Tolerancing―Tolerances of Form, Orientation, Location and Run-out

17. ISO 129-1/Amd 1, Technical Product Documentation（TPD）―Presentation of Dimensions and Tolerances―Part 1：General Principles

18. ISO 1302：Geometrical Product Specifications（GPS）―Indication of Surface Texture in Technical Product Documentation

19. ISO 14405-1：Geometrical Product Specifications（GPS）―Dimensional Toler-

ancing—Part 1：Linear Sizes

20. ISO 14405-2：Geometrical Product Specifications（GPS）—Dimensional Tolerancing—Part 2：Dimensions Other than Linear or Angular Sizes

21. ISO 14405-3：Geometrical Product Specifications（GPS）—Dimensional Tolerancing—Part 3：Angular Sizes

22. ISO 21920-1：Geometrical Product Specifications（GPS）—Surface Texture：Profile—Part 1：Indication of Surface Texture

23. ISO 21920-1（邦訳版），製品の幾何特性仕様（GPS）—表面性状：輪郭曲線—第1部：表面性状の図示

24. ISO 21920-2：Geometrical Product Specifications（GPS）—Surface Texture：Profile—Part 2：Terms, Definitions and Surface Texture Parameters

25. ISO 21920-2（邦訳版），製品の幾何特性仕様（GPS）—表面性状：輪郭曲線—第2部：用語，定義，表面性状パラメータ

26. ISO 25178-1：Geometrical Product Specifications（GPS）—Surface Texture：Areal—Part 1：Indication of Surface Texture

27. ISO 2768-1：General Tolerances—Part 1：Tolerances for Linear and Angular Dimensions without Individual Tolerance Indications

28. ISO 3274：Geometrical Product Specifications（GPS）—Surface texture：Profile Method—Nominal Characteristics of Contact（Stylus）Instruments

29. ISO/IEC Guide99：International Vocabulary of Metrology—Basic and General Concepts and Associated Terms（VIM）

30. ASME B46.1：Surface Texture（Surface Roughness, Waviness, and Lay）

31. 青木保雄，『改訂精密測定1』，コロナ社，1957

32. 青木保雄，『改訂精密測定2』，コロナ社，1958

33. M. Born, E. Wolf（草川　徹訳），『光学の原理II（第7版）』，東海大学出版会，2006

34. 電子工学ポケットブック編集委員会編，『電子工学ポケットブック』，オーム社，1983

35. 高　偉，清水裕樹，羽根一博，祖山　均，足立幸志，『Bilingual edition 計測工学 Measurement and Instrumentation』，朝倉書店，2017

36. 茨木創一，『工作機械の空間精度—3次元運動誤差の幾何学モデル・補正・測定—』，森北出版，2017

37. 情報通信研究機構（NICT），通信総合研究所季報 時間・周波数標準特集，**49**，2003

38. 株式会社ミツトヨ，精密測定機器の豆知識，カタログNo.11003(11)，2021

39. 金谷健一, 菅谷保之, 金澤 靖, 『3 次元コンピュータビジョン計算ハンドブック』, 森北出版, 2016

40. 香住浩伸, 『全図解 やさしい測定学』, 科学図書出版, 2002

41. 河田 聡編, 『超解像の光学』, 学会出版センター, 1999

42. 近藤雄基, 沼田宗敏, 吉田一朗, 『表面性状用ローパスフィルタの数理』, 東京図書出版, 2023

43. 公益社団法人精密工学会編, 『はじめての精密工学 第 1 巻』, 近代科学社, 2022

44. 公益社団法人精密工学会編, 『はじめての精密工学 第 2 巻』, 近代科学社, 2022

45. 公益社団法人精密工学会編, 『はじめての精密工学 第 3 巻』, 近代科学社, 2023

46. 公益社団法人精密工学会編, 『はじめての精密工学 第 4 巻』, 近代科学社, 2023

47. 工業技術院計量研究所編, 『計量技術ハンドブック』, コロナ社, 1987

48. 河野嗣男, 吉田嘉太郎編, 『インプロセス計測・制御・加工—機械加工の高度化を目指して』, 日刊工業新聞社, 1997

49. 桑田浩志編, 『ISO/JIS 準拠 製品の幾何特性仕様 GPS 幾何公差, 表面性状及び検証方法—ものづくりのデジタル化を進めるために』, 日本規格協会, 2012

50. 眞島正市, 磯部 孝, 『計測法概論 (上巻)』, コロナ社, 1970

51. 益田 正, 特異な量 "角度" とその標準, 精密工学会誌, **71**, 562-567, 2005

52. 森下光之助, 『機械学習を解釈する技術—予測力と説明力を両立する実践テクニック』, 技術評論社, 2021

53. 奈良治郎, 『表面粗さの測定・評価法』, 総合技術センター, 1983

54. 岡谷貴之, 『深層学習 改訂第 2 版 (機械学習プロフェッショナルシリーズ)』, 講談社, 2022

55. 斎藤康毅, 『ゼロから作る Deep Learning—Python で学ぶディープラーニングの理論と実装』, オライリージャパン, 2016

56. 産業技術総合研究所計量標準総合センター (NMIJ), 国際単位系 (SI) 基本単位の定義改定と計量標準, 2020

57. 産業技術総合研究所計量標準総合センター (NMIJ), 国際単位系 (SI) 第 9 版 (2019) 日本語版, 2020

58. 佐藤 淳, 『コンピュータビジョン—視覚の幾何学』, コロナ社, 1999

59. 社団法人精機学会 光を用いた工学表面の性状評価分科会, 『非接触方式による精密加工表面の性状評価—基礎と光学方式の応用—』, 理工企画社, 1985

60. 渋谷眞人, 大木裕史, 『回折と結像の光学』, 朝倉書店, 2005

61. 副島吉雄, 米持政忠, 『精密測定』, 共立出版, 1987

62. 曽我部東馬, 曽我部 完, 『Python による異常検知』, オーム社, 2021

63. 須賀信夫, 『精密測定入門』, ミツトヨ計測学院, 2014

64. 塚田忠夫，『機械設計工学の基礎―原理の理解から論理の展開まで』，数理工学社，2008

65. 津村喜代治，『基礎 精密測定』，共立出版，2005

66. 上野　滋，『はじめての計測技術』，大河出版，2016

67. 臼田　孝，『新しい1キログラムの測り方』，講談社，2018

68. 山本雄二，兼田二宏，『トライボロジー』，理工学社，2013

69. 谷田貝豊彦，『応用光学―光計測入門』，丸善出版，2005

70. 吉澤　徹，瀬田勝男編著，『光ヘテロダイン技術』，新技術コミュニケーションズ，1994

71. E. Abbe, Messgeräte für Physiker, *Zeitschrift für Instrumentenkunde*, **10**, 446-448, 1890

72. E. Abbe, On the Estimation of Aperture in the Microscope, *Journal of the Royal Microscopical Society*, **1**, 388-423, 1881

73. T. Beckwith, R. Marangoni, "*Mechanical Measurements*", Prentice Hall, 2006

74. G. Binnig, C. F. Quate, Ch. Gerber, Atomic Force Microscope, *Physical Review Letters*, **56**(9), 930-933, 1986

75. G. Binnig, H. Rohrer, Ch. Gerber, E. Weibel, Surface Studies by Scanning Tunneling Microscopy, *Physical Review Letters*, **49**(1), 57-61, 1982

76. BIPM, "*The International System of Units, 9th Edition (2019), (updated in 2022)*", 2022

77. BIPM, Recommended Values of Standard Frequencies

78. E. O. Brigham, "*The Fast Fourier Transform: An Introduction to Its Theory and Application*", Prentice Hall, 1988

79. The Consultative Committee for Length (CCL) of the International Committee for Weights and Measures (CIPM), *Mise en Pratique* for the Definition of the Metre in the SI, "*SI Brochure 9th edition (2019)*" Appendix 2, 2019

80. T. R. Corle, G. S. Kino, "*Confocal Scanning Optical Microscopy and Related Imaging Systems*", Academic Press, 1996

81. M. A. Curtis, F. T. Farago, "*Handbook of Dimensional Measurement*", Industrial Press, 2007

82. R. S. Figliola, D. E. Beasley, "*Theory and Design for Mechanical Measurements*", John Wiley & Sons, 2019

83. W. Gao (Ed.), "*Metrology (Precision Manufacturing)*", Springer, 2019

84. W. Gao, "*Precision Nanometrology: Sensors and Measuring Systems for Nanomanufacturing*", Springer, 2010

85. W. Gao, "*Surface Metrology for Micro- and Nanofabrication*", Elsevier, 2020

86. W. Gao, S. Goto, K. Hosobuchi, S. Ito, Y. Shimizu, A Noncontact Scanning Electrostatic Force Microscope for Surface Profile Measurement, *CIRP Annals-Manufacturing Technology*, **61**, 471-474, 2012

87. W. Gao, Y. Shimizu, "*Optical Metrology for Precision Engineering*", De Gruyter, 2022

88. J. Gawlikowski, *et al.*, Survey of Uncertainty in Deep Neural Networks, arXiv, 2107.03342, 2021

89. E. P. Goodwin, J. C. Wyant, "*Field Guide to Interferometric Optical Testing*", SPIE Press, 2006

90. R. Harlow, C. Dotson, "*Fundamentals of Dimensional Metrology*", Delmar Cengage Learning, 1989

91. E. Hecht, "*Optics*", Pearson, 2015

92. JCGM, "*JCGM 100, Evaluation of Measurement Data—Guide to the Expression of Uncertainty in Measurement*", 2008

93. JCGM, International Vocabulary of Metrology

94. D. J. Jones, *et al.*, Carrier-Envelope Phase Control of Femtosecond Mode-Locked Lasers and Direct Optical Frequency Synthesis, *Science*, **288**, 635-639, 2000

95. J. D. Kelleher, B. Mac Namee, A. D'Arcy, "*Fundamentals of Machine Learning for Predictive Data Analytics*", MIT Press, 2015

96. R. Leach (Ed.), "*Optical Measurement of Surface Topography*", Springer, 2011

97. Mitutoyo Corporation, Quick Guide to Precision Measuring Instruments, No. E11003(6), 2021

98. W. R. Moore, "*Foundations of Mechanical Accuracy*", The Moore Special Tool Company, 1970

99. A. C. Müller, S. Guido, "*Introduction to Machine Learning with Python*", O'Reilly Media Inc., 2016

100. M. S. Srinivasa, V. S. Murti, Scanning Probe Microscopy, *IEEE Control Systems Magazine*, **28**(2), 65-83, 2008

101. T. Udem, J. Reichert, R. Holzwarth, T. W. Hänsch, Accurate Measurement of Large Optical Frequency Differences with a Mode-Locked Laser, *Optics Letters*, **24**, 881, 1999

102. D. J. Whitehouse, "*Handbook of Surface Metrology*", Routledge, 1994

103. T. Yoshizawa, "*Handbook of Optical Metrology: Principles and Applications, Second Edition*", CRC Press, 2015

演習問題解答
Answers to Selected Problems

1-1） キログラム（記号は kg）は質量の SI 単位であり，プランク定数 h を単位 J s（kg m^2 s^{-1} に等しい）で表したときに，その数値を $6.626\,070\,15 \times 10^{-34}$ と定めることによって定義される．ここで，メートルおよび秒は c および $\Delta\nu_{Cs}$ に関連して定義される．

（The kilogram, symbol kg, is the SI unit of mass. It is defined by taking the fixed numerical value of the Planck constant, h, to be $6.626\,070\,15 \times 10^{-34}$ when expressed in the unit J s, which is equal to kg m^2 s^{-1}, where the metre and the second are defined in terms of c and $\Delta\nu_{Cs}$.)

1-2） 産業技術総合研究所計量標準総合センター（National Institute of Advanced Industrial Science and Technology, National Metrology Institute of Japan（NMIJ））

1-3） 温度（temperature）

1-4） $473.612\,7$ THz, 1.5×10^{-6}

1-5） $p(T)V = nN_A kT$, p：圧力（pressure），T：熱力学温度（thermodynamic temperature），V：気体の体積（volume of gas），n：気体の物質量（amount of gas），N_A：アボガドロ定数（Avogadro constant），k：ボルツマン定数（Boltzmann constant）．

2-1） 例えば，$S_V = 1.95$ mm, $N = 20$.（For example, $S_V = 1.95$ mm, $N = 20$.）

2-2） 例えば，文字盤 1 周 1 mm 相当として，$\theta_s = 3.6°$, $(N_2, r_2) = (18, 2.86)$, $(N_3, r_3) = (324, 51.57)$, $(N_4, r_4) = (18, 2.86)$.（For example, assuming that a 360° rotation of the pointer is equivalent to 1 mm, $\theta_s = 3.6°$, $(N_2, r_2) = (18, 2.86)$, $(N_3, r_3) = (324, 51.57)$, $(N_4, r_4) = (18, 2.86)$.)

2-3） $S_T = 0.01$ mm

2-4） 500 分割（500 divisions）

3-1） 厚さ 46.947 mm となるブロックゲージの組み合わせ（例えば，1.007 mm, 1.04 mm, 1.4 mm, 1.5 mm, 7 mm, 10 mm, 25 mm）（A combination achieving a thickness of 46.947 mm (for example, 1.007 mm, 1.04 mm, 1.4 mm, 1.5 mm, 7 mm, 10 mm, 25 mm)）

3-2） 少なくとも 26 ビット以上（More than or equal to 26 bits）

3-3) 9.70 nm

4-1) 結晶の形状に依存して変化する．結晶の幅を w，厚さを t とすると，AT カット
と GT カットの共振周波数はそれぞれ 160 490 820/t [kHz] と 329/w [kHz] となる．
(It varies depending on the shape of the crystal. If the width and thickness of the
crystal are w and t, respectively, The resonant frequencies of the AT cut and GT
cut are 166 490 820/t [kHz] and 329/w [kHz], respectively.)

4-2) 9 192 631 770 Hz

4-3) $v_{out} = v_{in} e^{-t/CR}$ となるので，時間と共に指数関数的に減少する．($v_{out} = v_{in} e^{-t/CR}$
and thus decreases exponentially with time.)

4-4) バンドパス・フィルタ (BPF) とは，特定の範囲の周波数だけを通過させるフィ
ルタのことである．具体的には，所望の周波数に対して低域の周波数を遮断するハイ
パスフィルタと高域の周波数を遮断するハイパスフィルタを組み合わせることで特定
の周波数の信号のみをえることができる．この所望の周波数を掃引することにより
RF スペクトルを得ることができる．

FFT を用いた方法は，時間的に変化する波形をデジタルストレージなどで記録し，
時間に対してフーリエ変換を行うことで周波数スペクトルを得る方法である．

(A band pass filter (BPF) is a filter that allows only a specific range of frequencies to pass through. Specifically, by combining a high-pass filter that blocks low
frequencies and a high-pass filter that blocks high frequencies with respect to the
desired frequency, it is possible to obtain only signals at a specific frequency. By
sweeping the desired frequency, the RF spectrum can be obtained.

The method using FFT is to record time-varying waveforms with digital storage
and perform Fourier transform with respect to time to obtain a frequency spectrum.)

5-1) 真円度とは，理論的に正確な円として与えられる線形体からの実形体の狂いの
大きさのことである．真円度によって定義される公差域は，対象の断面において，公
差 t の半径差を持つ同心の二つの円によって規制される．言い換えれば，真円度とは
「真円度測定機などから得られた測定データが，t mm（もしくは t μm）だけ離れた同
心の二つの円の内側に入らなければならない」ことを意味する．(Roundness is the
form error deviation of the real feature from the nominal toleranced feature that is
explicitly given as a circular line. The tolerance zone defined by the roundness, in
the considered cross-section, is limited by two concentric circles with a difference

in radii of *t*. In other words, the meaning of the roundness is that "an extracted circumferential line shall be contained between two circles separated by *t* mm (or *t* μm).")

5-2) プログラムコードの解答例として、MATLAB code で以下に示す。(An example of program code is shown below in MATLAB code.)

```
x=[1 2 3 4 5 6 7 8 9 10];
z=[4 8 7 3 1 6 9 4 9 7];

N = length(z);
n_w = 5;
N_shift = 1 + (n_w - 1)/2;
N_zma = N-(n_w - 1);

W = ones(1,n_w);
ZMA_sum = zeros(N_zma,1);

for i = 1 : N_zma
    for j = 1 : n_w
        ZMA_sum (i) = ZMA_sum (i) + z(i+j-1) * W(j);
    end
end

ZMA = ZMA_sum / n_w;

figure()
plot(x, z, '-ob','MarkerFaceColor','blue');
hold on ;
plot(x(N_shift :N-(N_shift - 1)), ZMA(1:N-(N_shift + 1)),'-or',
'MarkerFaceColor','white');
hold off ;
```

5-3) $0.000\,137\,\text{mm}\,(0.137\,\text{nm})@3°\,(\pi/60)$, $0.003\,53\,\mu\text{m}\,(3.528\,\text{nm})@15°\,(\pi/12)$

6-1) 線幅 1 pm のとき 0.4 m, 線幅 1 nm のとき 0.4 mm. (0.4 m for linewidth of 1 pm, 0.4 mm for linewidth of 1 nm.)

6-3) $200.000\,113\,\text{mm}$

6-4) 平行にすると凹部と凸部の違いがわからなくなるため. (Aligning them parallel would make it difficult to distinguish the differences between concave and convex areas.)

7-1) カメラ座標 x_L と x_R はそれぞれ式(A7-1), (A7-2)のように表せる. 式(A7-1), (A7-2)を式(7-3)〜(7-5)に代入すると, 式(8-7)〜(8-9)が得られる. (The camera coordinates x_L and x_R can be expressed as Eqs. (A7-1) and (A7-2), respectively. Substituting Eqs. (A7-1) and (A7-2) into Eqs. (7-3)-(7-5), Eqs. (8-7)-(8-9) are obtained.)

$$x_L = \frac{f}{\tan(\phi_1)} \qquad \text{(A7-1)} \qquad\qquad x_R = \frac{f}{\tan(\phi_2)} \qquad \text{(A7-2)}$$

7-2) 式(7-23)に式(7-24)〜(7-26)を代入すると, 左カメラと右カメラに関するカメラ行列が得られる. 得られたカメラ行列の要素を式(7-12)〜(7-15)に代入すると, 式(7-3)〜(7-5)が得られる. (Substituting Eqs (7-24)-(7-26) into Eqs. (7-23), we obtain the camera matrices for the left and right cameras. Substituting the elements of the obtained camera matrices into Eqs. (7-12)-(7-15), Eqs. (7-3)-(7-5) are obtained.)

7-3) $(X_w, Y_w, Z_w) = (70, 20, 250)$ が得られる. 行列を用いて式(7-12)〜(7-15)を式(A7-3)の形式にまとめると, 最小二乗解は式(A7-4)を用いて求められる. ただし, *inv* は逆行列を表す. 行列 A, B はカメラ行列の要素と画像座標から求められる. ($(X_w, Y_w, Z_w) = (70, 20, 250)$. Using matrices, Eqs. (7-12)-(7-15) can be summarized in the form of Eq. (A7-3), and the least-squares solution can be obtained using Eq. (A7-4). Note that *inv* denotes the inverse matrix. The matrices A and B are obtained from the elements of the camera matrix and the image coordinates.)

$$A \begin{bmatrix} X_w \\ Y_w \\ Z_w \end{bmatrix} = B \qquad \text{(A7-3)} \qquad\qquad \begin{bmatrix} X_w \\ Y_w \\ Z_w \end{bmatrix} = inv(A^T A) A^T B \quad \text{(A7-4)}$$

7-4) 直接 TOF 法であれば時間の計測誤差, 間接 TOF 法であれば受光量の計測誤差などによる, t_d の計測誤差. 多重反射光, 散乱光など, 想定した光路と異なる光路からの光の影響. (The measurement error of t_d due to the measurement error of time in the case of the direct TOF method, or measurement error of the amount of received light in the case of the indirect TOF. The influence of light from different light paths than expected, such as multiple reflections and scattered light.)

8-1) 例えばレーザ測長機とプローブの位置にオフセットが生じない構造とする. 現実的には計測対象物との干渉があるため, 完全にアッベ誤差を生じなくさせる構造を実現するためには工夫が必要である. (The structure should be such that no offset exists between the position of the position (such as a laser interferometer) and the probe. In reality, an ingenious design is required to completely eliminate the Abbe error due to interference with the object to be measured.)

8-2) 不確かさの伝播則より，合成標準不確かさ u_c は式(A8-1)のとおり求められる．(From the uncertainty propagation law, the combined standard uncertainty u_c can be obtained in Eq. (A8-1).

$$u_c = \sqrt{\left(\frac{\partial p_x}{\partial \phi} u(\phi) \right)^2} = \sqrt{(-l \times u(\phi) \times \sin(\phi))^2} \tag{A8-1}$$

8-3) 不確かさの伝播則より，合成標準不確かさ u_c は式(A8-2)のとおり求められる．(From the uncertainty propagation law, the combined standard uncertainty u_c is obtained in Eq. (A8-2).)

$$u_c = \sqrt{\left(\frac{\partial p_x}{\partial \phi_1} u(\phi_1) \right)^2 + \left(\frac{\partial p_x}{\partial \phi_2} u(\phi_2) \right)^2} \tag{A8-2}$$

8-4) 式(8-15)〜(8-17)の連立方程式を解いて，$(p_x, p_y, p_z) = (50\,\text{mm},\ 50\,\text{mm},\ 70.7\,\text{mm})$が得られる．(Solving the simultaneous Eqs. (8-15)-(8-17), we obtain $(p_x, p_y, p_z) = (50\,\text{mm}, 50\,\text{mm}, 70.7\,\text{mm})$.)

9-1) 理想的な2つの点光源を結像した際，それぞれの回折像が生じる．それら2つの回折像が重なり合ったとき，中間地点が回折像の明るさと同じであれば，その2点は区別できるというもの．$\alpha=0.51$．(When two ideal point light sources are imaged, diffracted images of each are generated. When these two diffracted images are superimposed, the two points can be distinguished if the brightness of the intermediate point is the same as the diffracted image. $\alpha=0.51$.)

9-2) 0.82

9-3) 回折限界は 540.9 nm，焦点深度は 1 478 nm．(540.9 nm for diffraction limit, and 1 478 nm for depth of focus.)

9-4) 例えば，凹レンズと凸レンズを用いて補正することができる．(As one representing method, concave and convex lenses are used to correct the spherical aberration.)

9-5) 面形状の変化が大きいと面内方向のインターフェログラム変化が急峻となり，白色干渉計の横分解能より狭い領域で変化が生じると，インターフェログラム変化を捉えられないため．(The interferogram change in the in-plane direction becomes steeper when the change in surface shape is large. If the change is steeper than the lateral resolution of the white light interferometer, the interferogram change cannot be measured.)

10-1) 原子や分子の励起, 分子の結合の切断, 原子の操作. (Excitation of atoms and molecules, cleavage of molecular bonds, atom manipulation.)

10-2) $Z=2^{\frac{1}{6}}\sigma$

10-3) 例えば, AFM カンチレバーや STM プローブの側面がプローブ先端より先に試料と接触すると, プローブ先端は試料表面近傍まで接近できないため測定できない. (For example, if the side wall of the AFM cantilever or STM probe comes into contact with the sample surface before the tip of the probe, the tip of the probe cannot approach the vicinity of the sample surface. Consequently, the measurement become impossible.)

11-1) $\delta_{\cos}=1000(1-\cos 1°)=0.1523$ mm

11-2) $\delta_{\mathrm{Abbe}}=100\times\tan 1°=1.7455$ mm

11-3) $1000\{1+1.2\times10^{-6}\times(23-20)\}=(R'-0)\times\{1+17.3\times10^{-6}\times(23-20)\}$

$R'=1000\cdot\dfrac{1+1.2\times10^{-6}\times(23-20)}{1+17.3\times10^{-6}\times(23-20)}=999.951$ mm

11-4) $s=1500\times0.2133=316.95$ mm

11-5) ダイヤモンド (diamond)：$E_1=1050$ GPA, $\nu_1=0.1$

アルミニウム合金 A2017 (aluminum-alloy A2017)：$E_2=72.6$ GPa, $\nu_2=0.33$

$$\frac{1}{E^*}=\frac{1-0.33^2}{72.6\times10^9}+\frac{1-0.1^2}{1050\times10^9}$$

$E^*=75.66\times10^9$ Pa

$$\delta=\left(\frac{9\times(0.1\times10^{-3})^2}{16\times(75.66\times10^9)^2\times5\times10^{-6}}\right)^{\frac{1}{3}}=0.00581\ \mu\mathrm{m}$$

$$P_{\mathrm{mean}}=\frac{1}{\pi}\left(\frac{16\times(0.1\times10^{-3})\times(75.66\times10^9)^2}{9\times(5\times10^{-6})^2}\right)^{\frac{1}{3}}=1.095\times10^9\ \mathrm{Pa}=1095\ \mathrm{MPa}$$

P_{mean} が A2017 の 0.2% 耐力 (275 MPa) の 3 倍より大きいため塑性変形が生じる. (The aluminum-alloy surface will be damaged due to plastic deformation since P_{mean} is larger than 3 times of the 0.2% offset yield strength of the aluminum-alloy (275 MPa).)

12-1) $I_0(x,y)=\dfrac{4\pi}{\lambda}\left[e_Z(x,y)-a(x,y)\right]$

$$I_{X+1}(x,y)=\frac{2\pi}{g}e_X\left(\frac{x}{\cos(\theta/2)},y\right)+\frac{4\pi}{\lambda}\left[e_Z\left(\frac{x}{\cos(\theta/2)},y\right)\cos\frac{\theta}{2}-a(x,y)\right]$$

$$I_{X-1}(x, y) = -\frac{2\pi}{g} e_X\left(\frac{x}{\cos(\theta/2)}, y\right) + \frac{4\pi}{\lambda}\left[e_Z\left(\frac{x}{\cos(\theta/2)}, y\right)\cos\frac{\theta}{2} - a(x, y)\right]$$

12-2) $m_1(\theta) = r(\theta) + s_1(\theta)$

$m_2(\theta) = r(\theta) - s_2(\theta)$

$$\frac{m_1(\theta) + m_2(\theta)}{2} = r(\theta) + \Delta r(\theta)$$

$$\Delta r(\theta) = \frac{s_1(\theta) - s_2(\theta)}{2}$$

12-3) $r(\theta) = A\cos(N\theta)$

$\Delta\phi = 2\pi/N$

$m_i(\theta) = r(\theta - i\Delta\phi) + s_i(\theta) = A\cos(N\theta - i2\pi) + s(\theta) = A\sin(N\theta) + s(\theta)$

$$\bar{m}(\theta) = \frac{1}{N}\sum_{i=0}^{N-1} m_i(\theta) = A\sin(N\theta) + s(\theta) = r(\theta) + s(\theta)$$

$r(\theta) = m_0(\theta) - s(\theta) = 0$

12-4) $\phi = 10°$, $\tau = 30°$ の場合に比べて，$\phi = 10°$，$\tau = 34°$ の場合は非対称的なので，感度が0になる最初の調和次数は後者の方が高い．(Compared with the case of $\phi = 10°$, $\tau = 30°$, the probe arrangement is more assymmetric in the case of $\phi = 10°$, $\tau = 34°$. Therefore, the latter has a higher first harmonic with zero sensitivity.)

12-5) $\Delta f(x) = \dfrac{\alpha}{2d^2}\left(\dfrac{L}{2}\right)^2 = 4.5\,\mu\mathrm{m}$

13-1) オートエンコーダとは，エンコーダとデコーダを組み合わせたアルゴリズムである．エンコーダでは，入力値の次元を徐々に少なくし（圧縮し），デコーダでは，エンコーダで圧縮した次元を元に戻すことを行う．エンコーダ時に，特徴量のみを抽出する作業を行うことから，オートエンコーダを用いることで入力値のノイズを除去する効果が期待できる．

　一方で，最小二乗法では，ノイズも含まれた一連のデータ群を理論値に基づいて構築された数式中の係数の最良推定値を求める方法である．具体的には，理論値と実験値との差分が最小になるように係数を決定する．ランダムに生じる特定のノイズ群が含まれていたとしても，ノイズの影響を排除して解析を行うことができる．

　いずれの方法も，効果の面では，入力値からノイズを除去できるという点で同じである．

　(An autoencoder is an algorithm that combines an encoder and a decoder. The encoder gradually reduces (compresses) the dimensionality of the input values, and the decoder restores the dimensions compressed by the encoder. Since only the feature values are extracted during encoding, the autoencoder is expected to be effective in removing noise from the input values.

On the other hand, the least-squares method is a method for obtaining the best estimate of the coefficients in a mathematical expression constructed based on theoretical values from a set of data that also includes noise. Specifically, the coefficients are determined so that the difference between the theoretical and experimental values is minimized. Even if a particular group of randomly generated noise is included, the effect of the noise can be eliminated from the analysis.

Both methods have the same effect of removing noise from the input values.）

13-2) U-Net は，複数の畳み込み層を並列に処理する方法の1つである．この複数の畳み込み層は，畳み込む範囲が異なるものを組み合わせるのが一般的である．したがって，例えば，画像処理では，ボケた画像にノイズが重畳された状況が想定される．ボケを除去するためには局所的な影響のみに効果を発揮する必要があるが，ノイズ処理を直列的に処理してしまうと，ボケを除去するプロセスに影響を与えてしまう．このような場合は，並列に処理する必要があり U-Net を使うと効果的である．

ResNet は，複数の畳み込み層で構成されたアルゴリズムで，いくつかの層をスキップするプロセスが組み込まれたアルゴリズムである．一般的に，層の数を増やすと様々な特徴量が抽出できる代わりに，情報量が著しく少なくなるためにある程度の層までしか増やすことができない．しかし，スキップする層を付加すると層の数をかなり増やすことができるため情報量の欠落を減らしつつ，多数の特徴量を抽出することができるようになる．

（U-Net is a method for processing multiple convolutional layers in parallel. The convolutional layers are generally combined with different convolutional ranges. Thus, for example, in image processing, a situation in which noise is superimposed on a blurred image is assumed. To remove the bokeh, it is necessary to effect only local effects, but if the noise processing is done serially, it will affect the process of removing the bokeh. In such cases, parallel processing is necessary and effective using U-Net.

ResNet is an algorithm that consists of multiple convolutional layers and incorporates the process of skipping some layers. In general, increasing the number of layers enables the extraction of a variety of features, but the amount of information is significantly reduced, so the number of layers can only be increased to a certain extent. However, the addition of skipped layers can considerably increase the number of layers, thus enabling the extraction of a large number of features while reducing the amount of missing information.)

13-3) 活性化関数を線形変換にすると，出力値に係数を乗算するだけになるので層の数を増やすことにならない．(If the activation function is a linear transformation, the number of layers is not increased because the output values are simply multiplied by the coefficients.)

13-4) 学習させすぎると，本来のノイズのような本来のシグナルとは関係のない変化にも追従するようにアルゴリズム中の係数を最適化してしまう．したがって，学習した信号とは異なる信号を入力すると真値に対して大きく異なる推定値が導出されてしまう．(Over-learning optimizes the coefficients in the algorithm to follow changes that have nothing to do with the original signal, such as the original noise. Therefore, if a signal different from the learned signal is input, estimates that differ significantly from the true value will be derived.)

14-1) 2 m

14-2) 25 THz

14-3) 192 THz, 4.93 THz

14-4) $v_c = \dfrac{\omega}{k}$, $v_g = \dfrac{d\omega}{dk}$

14-5) $G_P(\omega) = \displaystyle\int_{-\infty}^{\infty} g_P(t) e^{-j\omega t} dt$

$E_P(t) = g_P(t) E_c(t) = g_P(t) e^{j\omega_c t}$

$H_P(t) = \displaystyle\int_{-\infty}^{\infty} E_P(t) e^{-j\omega t} dt = \int_{-\infty}^{\infty} g_P(t) e^{j\omega_c t} e^{-j\omega t} dt = \int_{-\infty}^{\infty} g_P(t) e^{-j(\omega - \omega_c)t} dt = G_P(\omega - \omega_c)$

和文索引
Index in Japanese

欧文索引
Index in English

Bilingual edition
精密計測学
Precision Metrology

定価はカバーに表示

2024 年 4 月 5 日　初版第 1 刷

著　者　高　　　　偉
清　水　裕　樹
水　谷　康　弘
道　畑　正　岐
河　野　大　輔
吉　田　一　朗
伊　東　　　聡
清　水　浩　貴
発行者　朝　倉　誠　造
発行所　株式会社　朝　倉　書　店
東京都新宿区新小川町 6-29
郵便番号　162-8707
電　話　03(3260)0141
FAX　03(3260)0180
https://www.asakura.co.jp

〈検印省略〉

中央印刷・渡辺製本

ISBN 978-4-254-20178-9　C 3050

Printed in Japan

東北大 高　偉・北大 清水裕樹・東北大 羽根一博・
東北大 祖山　均・東北大 足立幸志著

Bilingual edition 計測工学 Measurement and Instrumentation

20165-9 C3050　　　　　　A 5 判 200頁 本体2800円

計測工学の基礎を日本語と英語で記述。〔内容〕計測の概念／計測システムの構成と特性／計測の不確かさ／信号の変換／データ処理／変位と変形／速度と加速度／力とトルク／材料物性値／流体／温度と湿度／光／電気磁気／計測回路

神野郁夫・小寺秀俊・鈴木亮輔・田中　功・
冨井洋一・中部主敬・箕島弘二・横小路泰義著

計　　測　　工　　学

20159-8 C3050　　　　　　A 5 判 192頁 本体2300円

測定の実際にあたっての基礎事項を丁寧に解説。さらに機械，材料，原子核，エネルギー，物理工学などの専門分野での測定方法と先端技術までをまとめた。計測器の根本原理や，開発の契機なども述べ，より興味がわくようまとめた。

東洋大 窪田佳寛・東洋大 吉野　隆・東洋大 望月　修著

きづく！つながる！ 機 械 工 学

23145-8 C3053　　　　　　A 5 判 164頁 本体2500円

機械工学の教科書。情報科学・計測工学・最適化も含み，広く学べる。〔内容〕運動／エネルギー・仕事／熱／風と水流／物体周りの流れ／微小世界での運動／流れの力を制御／ネットワーク／情報の活用／構造体の強さ／工場の流れ，等

松井　悟・竹之内和樹・藤　智亮・森山茂章著

初めて学ぶ図学と製図 改訂版

23152-6　　　　　　　　　A 5 判 192頁 本体3300円

図形に関する基礎力が習得できるようやさしく解説。〔内容〕投影／点・直線・平面の投影図／副投影法／回転法／切断／展開／イラストレーション／図面の基本／図形の表し方／断面図／寸法記入法／スケッチ／機械要素の製図法／他

早大 富岡　淳編著

役 に た つ 機 械 製 図 （第3版）

23142-7 C3053　　　　　　B 5 判 272頁 本体4300円

自動車工学の基礎知識に加え，自動運転の基盤となる運転操作のモデル化や，車両運動制御の基礎と応用を解説。〔内容〕タイヤの力学モデル／制駆動系の力学モデル／クルマの運動モデル／ドライバ操作のモデリング／ヒトの運動感受性／他

滋賀大 竹村彰通監訳

機　　械　　学　　習
―データを読み解くアルゴリズムの技法―

12218-3 C3034　　　　　　A 5 判 392頁 本体6200円

機械学習の主要なアルゴリズムを取り上げ，特徴量・タスク・モデルに着目して論理的基礎から実装までを平易に紹介。〔内容〕二値分類／教師なし学習／木モデル／ルールモデル／線形モデル／距離ベースモデル／確率モデル／特徴量／他

宇都宮大 谷田貝豊彦著

光　　　　　　　　　学

13121-5 C3042　　　　　　A 5 判 372頁 本体6400円

丁寧な数式展開と豊富な図解で光学理論全般を解説。例題・解答を含む座右の教科書。〔内容〕幾何光学／波動と屈折・反射／偏向／干渉／回折／フーリエ光学／物質と光／発光・受光／散乱・吸収／結晶中の光／ガウスビーム／測光・測色／他

渡辺敏行著

材料・化学系技術者のための光学入門

13148-2 C3042　　　　　　A 5 判 264頁 本体4000円

化学系学生や材料開発に携わる技術者のための入門書。〔内容〕ベクトル解析／マクスウェル方程式・波動方程式／光波の基礎／屈折率・異方性／物質と光の相互作用／偏光／反射・屈折／光ファイバー・光導波路／波動の重ね合わせ／回折／他

黒田和男・荒木敬介・大木裕史・武田光夫・
森　伸芳・谷田貝豊彦編

光 学 技 術 の 事 典

21041-5 C3550　　　　　　A 5 判 488頁 本体13000円

カメラやレーザーを始めとする種々の光学技術に関連する重要用語を約120取り上げ，エッセンスを簡潔・詳細に解説する。原理，設計，製造，検査，材料，素子，画像・信号処理，計測，測光測色，応用技術，最新技術，各種光学機器の仕組みほか，技術の全局面をカバー。技術者・研究者必備のレファレンス。〔内容〕近軸光学／レンズ設計／モールド／屈折率の計測／液晶／レーザー／固体撮像素子／物体認識／形状の計測／欠陥検査／眼の光学系／量子光学／内視鏡／顕微鏡／他

上記価格（税別）は 2024 年 2 月現在